Extreme Fundamentals of Technology

A Primer of Computers, Electronics, and Technology

Second Edition

By Bob Dukish

AF405240

Website: www.dukish.com

For information: bob@dukish.com

Published by Dukish LLC

ISBN 9798669211974

Warning:

      Electrical circuits and components may contain lethal voltages even when disconnected. Do not attempt to test, modify, or repair electrical equipment. Hazardous voltages are present when equipment covers are removed.

Preface

The original book, and now the second update, is an introductory guide to basic science and engineering concepts useful to people of all ages who are interested in technology. We originally developed the idea of Extreme Technology at an advisory committee meeting for a County Career and Technical Center, as a way to gain higher interest from prospective students to pursue a college-level engineering program of study. The underlying goal of Extreme Technology is to present the needed background information to students, but then allow them to explore subjects in science and technology in creative and innovative ways. The latest cutting-edge methodology in Education is to incorporate project and problem-based learning. We take the learning approach to a higher level by adding what we term as *predictable unpredictability* into the curriculum. In many current programs utilizing these methods, such as Project Lead the Way, the students are introduced to the related concepts and given a constrained set of exercises and materials for hands-on kinesthetic activities. They then progress through a predetermined lesson to reach a terminal performance outcome. We accomplish the same standards, but unconventionally, in a way that is far more interesting to both students and teachers. Extreme Technology Education may seem haphazard to the casual observer. Not all projects are hands-on because the technology may not yet be in existence, and not all problems can be solved because there may be a need for more scientific research. The Extreme Technology pedagogy allows the student to dream, invent, and think *laterally,* while applying known concepts to explore the unknown. In developmental psychology, it is believed young children go through a stage early in life where they exhibit animism as they see the world as somewhat magical. We want to allow our students to return to this way of thinking but supplied with the tools to make the magic more than just make-believe. The book is designed for both high school and adult learners to understand current technology, and to act as a reference for the exploration of futuristic technology.

Changes to the second edition

Time marches on, and the years flew by since the release of the first edition. More than ten years have elapsed, and the world has evolved to depend on more streamlined and powerful technological products led by the cellphone. Now with 5-G devices, the connectivity speeds will provide a conduit for much higher data flow and more processing within the cloud. Individual processors have gotten much faster and have become highly miniaturized for computers, cellphones, and Internet of Things (IoT) devices. The miniaturization is also leading to a wide variety of very inexpensive smart products. As devices become more functional and less expensive, they turn into somewhat of a commodity item and not given a second thought, just as one does not think much about the construction of the toaster when making breakfast. It is very gratifying to see the rise of the maker community growing in numbers and hacking and tinkering with new ideas. Much of the credit goes to the founders of the Arduino project; Massimo Banzi, David Cuartielles, David Mellis and others who developed an easy to understand programming language that evolved from the earlier platforms called *Wiring* and *Processing*. They incorporated an easy to use language with an excellent IDE for use with a low-cost processor, the Atmel ATmega, for programming an inexpensive open-source development board. Today the Arduino microcontroller community numbers in the millions.

The other contributor to the surge in the maker movement was the advent of the Raspberry Pi. It is beautiful for computer education in schools and uses as an extremely low-cost alternative to PCs for those in the developing world. The Raspbian operating system is very efficient, and the Pi's hardware makes it a valid substitute for a standard PC. I can see how a young person can develop an affinity to the Pi with the visibility of the components and connectors on the board. With its powerful multi-core processor and Internet connectivity, it runs beautifully as a computer. One of the project goals is to have the users get familiar with coding through the use of a modular method called *Scratch*, or with *HTML* webpage code, and ultimately with *Python* – one of today's most popular programming languages. Just as with Arduino aficionados, the Pi community has many Webpages and online resources available.

Both the Arduino and Raspberry Pi are reasons for a large amount of interest in the maker movement. Each has its purpose with some small amount of overlap. The Arduino is a microcontroller and can be used with many types of input sensors to operate devices such as lighting or sounding indicators, for controlling motor movements, or sending data via bus or Bluetooth. Both can and even work as an IoT device. The Raspberry Pi's purpose is more of what could be considered a standard computer. It additionally has the ability for I/O, and with its low-cost, the Pi is sometimes used for projects where it acts as a controller.

Going from a small number of people in the early computer days exchanging messages on Newsgroups through USENET, we now find almost all people in the developed world addicted to social messaging platforms like Facebook, Twitter, TikTok, etc. It is effortless to use the phone apps, and sometimes we use them at our peril with risks to our reputations. Just the other day, I made what I thought was a very humorous comment on a post by Massimo about the latest Arduino and got a response from him saying something to the effect, "Don't look at your nose, look at the Moon." I am now second-checking what I think is funny before posting to social media. Sorry, Massimo! (I'm a huge fan of Arduino.) I doubt if the Pandemic had increased social media usage because the saturation had to be nearly 100 percent before social distancing took effect. This past year I attended a party in a beautiful park on a warm autumn day. The leaves were at their peak colors, and the park was just extraordinary. I used my phone to snap a picture of a group of around ten friends. They weren't enjoying the park's beauty, nor were they talking or looking at each other; instead, they were each glued to their phone screens, either messaging or scrolling through social media posts. I think we have created a monster and a blight on society, but the technology behind the scenes is what is exciting.

There are vast amounts of products and applications for the use of basic technology, and this book is about technology's most fundamental aspects. If you are old enough, you may remember the child's toy called an erector set, and we all are most certainly familiar with Legos. This goal of this book is to provide a basic knowledge of a wide breadth of fundamental concepts in a similar way to understanding the pieces of an Erector or Lego set. A profound understanding of fundamental building blocks will help you to use your creativity to build marvelous devices. Additionally, the fundamental building blocks stay fairly consistent over a long time and is one of the reasons it was not necessary until now to update this book. Much of the text has only been expanded to reflect the underlying concepts of some of the popular products we see today. Software is ever-present in today's Maker Movement, and this edition delves deeper into that area. We also introduce more project applications along with the fundamental concepts on which they are built. We also have included chapter summaries so that the takeaway is more evident. Some of the information has been repositioned from the last edition to help with the flow of material. This edition has been more strictly edited to make some areas less ambiguous. I hope the content we cover will be helpful in your future endeavors, and also that this is a fun and interesting book.

6

Acknowledgments

I would like to thank my students. I learned a great deal from them over the years.

About the author

Bob Dukish has spent over 40 years working and teaching in the field of technology. After serving in the military, working as an electronics component engineer, and running a corporation, he now teaches at Kent State University. He has Associate Degrees in Avionic Systems, and Electronics Engineering Technology. A Bachelor's Degree in Physics from Syracuse University, as well as Master's degrees from Kent State University, and Rensselaer Polytechnic Institute. He earned his last degree at age 54, and he considers himself a lifelong learner.

Table of contents

Chapter 1
Dealing with numbers

Section 1.1. Normal size things

Mathematics is easy and fun for some people, but not for me. As a technologist, I consider working with numbers as a necessary evil. In designing a product or producing code, we look at the big-picture first and worry about the details later. You might sketch a design for a product or write a flow chart for coding a program as the first step in the design process. A popular 3-D CAD software package from Autodesk called Inventor can start with the user drawing a sketch and later adding values as constraints. The excitement of bringing a new design or code to reality usually will give the push to buckle-down and work the numbers. Sometimes our design process can include rounding off to some extent when working with large items, but there are other times when precision is necessary. An architect who designs skyscrapers has a different perception than a machinist working with small complex mechanical devices. The two professions even use different types of rulers. An architect's ruler makes it easy to scale-up blueprints to full-size dimensions, whereas a machinist's ruler has many more increments for the intricate measurement of small parts. It is usually best to represent things precisely with mathematics to avoid any confusion because of differing perceptions. The Hubble space telescope, shown in Figure 1.1, was built with some people using the metric system while others were using the English system. The first images it took were out of focus, but luckily astronauts were able to repair it while in orbit. It has been providing exquisite photos illustrating the grandeur of the universe ever since. Thank goodness for workarounds.

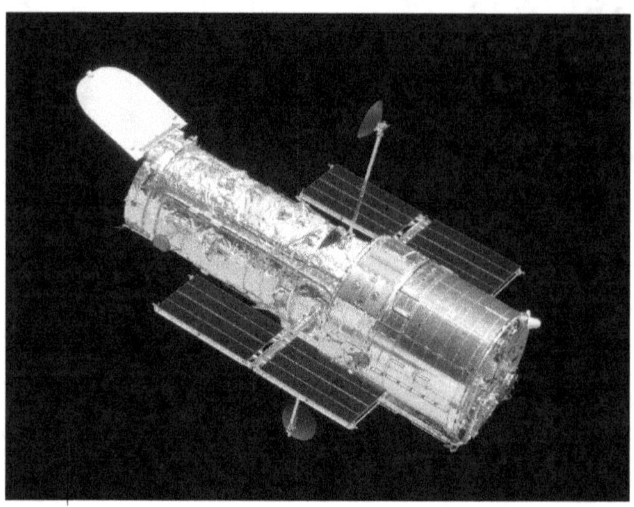

Figure 1.1. Nasa Hubble Telescope

Let's precisely analyze how we could mathematically describe objects of what would be considered normal size. The floor we walk on in a room could be called a plane in mathematics and is the subject of plane geometry. A typical math problem dealing with a plane might be to find the size of a carpet for the floor of a room. The problem is easy if you have a rectangle and know the length and width of the room. When you multiply those two numbers together, you find the rug area size A = L x W. As we move through a room, we might interact in three distinctly different ways. We can move side-to-side, front-to-back, and we also can move up-and-down. Another typical geometric math problem is to find the amount of air in a room, as shown in Figure 1.2. This problem involves all three directions. You could multiply the floor area by the height of the room to find the cubic volume, so V = L x W x H.

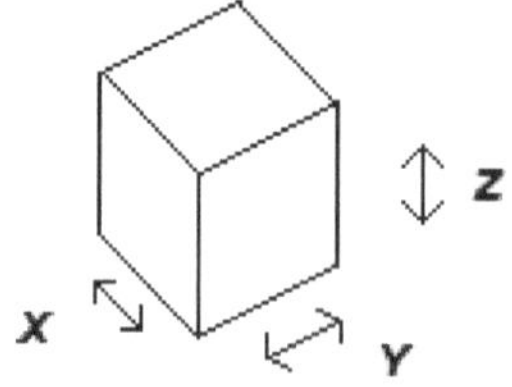

Figure 1.2. Cubic volume

Many times, the letter variables representing quantities in an algebra problem get confusing. Also, there are different ways to write formulas that mean the same thing. In the English language, we have words that are different but mean the same thing called synonyms. People that are skilled writers and speakers have an extensive vocabulary and can write and say things in different ways to tailor the delivery to a specific audience. With a little practice, you can build-up your math vocabulary, which will enable you to make your point in the most effective way.

<u>An example of why some people hate Math</u>

In our volume of air in a room problem, we could have written the formula in a few different but equal ways:

Rather than V = L x W x H, we might have said V = A x H because we earlier found the floor area, so A = L x W. Some people use a dot instead of the "x" symbol. That is slightly incorrect because a dot product has a slightly different meaning. The preferred way is to put the letters together to imply the multiplication function: V = L W H. Parentheses also are used to describe multiplication. V = (L)(W)(H) is correct but is overkill. It's like playing any game successfully; the first thing one needs to know are the rules of the game. There is a short refresher of the basic rules of Math in the appendix.

To find our way around two and three-dimensional space, the 17th-century French mathematician Rene Descartes explained it in a system we now call *Cartesian Coordinates*. The directions are represented with the following letters: "X" for the side-to-side horizontal direction, "Y" for the front-to-back direction, (the x and y form our plane), and we use "Z" for the up-and-down vertical direction. We can assign numerical values to the distance we move in each direction, and the three letters "X, Y, Z" can represent every unique point in the room, as shown in Figure 1.3.

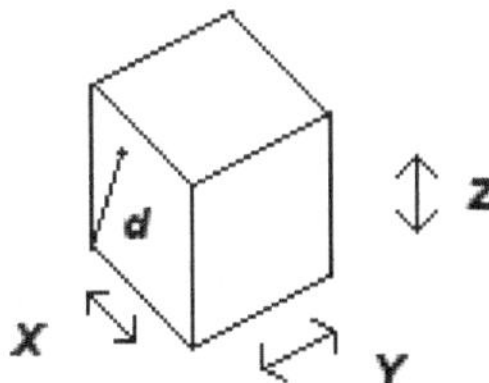

Figure 1.3. Location in three-dimensional space

To explain where we end up, we have to make it understood what our beginning point of origin was, and it is our point of reference. The origin will have the location (0,0,0), and every place we walk in the room could be represented by three numbers in the (X, Y, Z) format. It would tell how far we moved in each of the directions from the original starting point. We will use positives in the examples that follow, but –X would imply moving to the left of the origin, -Y would be backward, and –Z is below the floor.

A question might also arise about how far we moved in a straight line from the point of origin. An ancient Greek mathematician named Pythagoras came up with a formula for finding the length of a right triangle's hypotenuse. We will combine both Descartes and Pythagoras' work to quantify the position of objects in three dimensions. If the origin is at (0,0,0), take each X, Y, and Z number and square it, then add each of the resultant values together, and take that number's square root. This value gives the direct distance from the origin to the point where we end up. There are a couple of ways to solve a problem like that with a calculator. If you do it just like we said, you can avoid using parentheses on a calculator.

<u>Method of working a problem on a calculator without the use of parenthesis</u>

      Step 1. Enter the X number and square it.

      Step 2. Push the + button followed by the Y number and square it.

      Step 3. Push + again and enter the Z number squaring it too.

(That's not the answer yet!)

      Step 4. Press = and finally, the square root button $\sqrt{}$ for the final result.

The way just illustrated is sort of working the math problem from the inside out. Many people are comfortable using parentheses to determine the order of operation, but parentheses may become a burden if there are many groupings involved. In some problems, there might be parentheses inside of parentheses. In a problem like that, they are referred to as *nested*. You might try both ways to work a problem and see which you like best. Either method is fine, as long as you get the same result.

<u>Example of the Pythagorean Theorem using coordinates</u>

<u>Problem</u>

      If you stand in the center of the floor holding a ball in your hand with your arm down toward the side of your right leg, and we say the ball is at the origin (0, 0, 0). Then you walk across the room by going 10 units to the right in the X direction, and 10 units in the forward direction, and then raise the ball 2 units high, how far has the ball traveled? (Note: the formula below is used for three directions. The standard form for a plane does not have the z^2 term that we use for height.)

<u>Solution</u>

Using the Pythagorean Theorem:

$$d = \sqrt{x^2 + y^2 + z^2}$$

And

$$d = \sqrt{10^2 + 10^2 + 2^2} = 14.28 \text{ units.}$$

So, to go from the Cartesian Coordinates (0,0,0) to (10,10,2) is a change in direct distance of 14.28 units from the origin.

Tip: To enter parentheses in some calculators, it would look like this:

$$d = \sqrt{((10)^2 + (10)^2 + (2)^2)} = 14.28 \text{ units}$$

You could enter everything after the equal sign directly in the calculator going from left to right. The inside parentheses may not be necessary on most calculators, but entering them won't matter. Just remember that parentheses must be in a set, one to open the grouping and one to close it.

The material we have just covered is very applicable to CNC machining and 3-D printing. CNC (Computer Numerical Control) is an automated process of machining where the tooling can even be automatically changed as the CNC machine is running. CNC is considered *subtractive* manufacturing since holes may be drilled or material milled. 3-D printing is *additive* since the material is being deposited as the object is being printed. The first step in both methods is setting the zero reference point, sometimes referred to as *home*, and all movements are referenced to that point. There is low-cost equipment available for Makers interested in manufacturing using both methods. A controller board such as the Arduino can be used to both position and operate the equipment.

There is also a fourth dimension to our lives. Time is the fourth dimension, and since 3-D printing takes so long to complete, it really should be called 4-D. Jokes aside, you might be sitting in a chair in a room right now, but next hour someone else might be in the chair, and you may be in a different room. In some cases, you may need to give a position referenced to time. To make matters worse, some philosophers believe that there is no present time, only past and future. They feel that because no matter how short the time, the time has passed as soon as something occurs. No matter how small a number that you could measure, there would be one smaller. We'll get to really tiny things later, but now since we have been exploring a room's area, let's move on and explore really big things. In our three-dimensional universe (really four), the biggest thing is the entire universe. Because of its enormous size, there really isn't a practical way to move from one side of the universe to the other unless you can travel faster than light, and light is the fastest thing that exists under normal circumstances. You may have noticed how much faster it is than sound by observing a distant fireworks display or a summertime thunderstorm.

Section 1.2. Really big things

Scientists have approximated the visible universe's size by using the fact that all lighted objects in the universe have a Doppler shift. We can't experience the Doppler shift for light in our everyday lives, but you may have noticed it as a change in pitch as a train whistle or car horn sounds as it goes by you; the sound changes in tone from a higher to lower pitch as the object whizzes past. It may sound something like this: weeeeeeeee…ahhhhhhhhh… The sound is higher in pitch coming towards you, and then goes lower as it moves away. That is because as the sound is coming toward you, the waves get compressed. When the object is traveling away from you, the waves become elongated. Higher sounds have more tightly packed waves, and lower sound waves are stretched out. Astronomers noticed this same effect in the light from stars. The smaller light waves look blue to our eyes, whereas longer light waves appear red. All of the other colors of the rainbow are in-between. It was observed that stars and other objects in the universe exhibit a redshift in their light. Astronomers like Edwin Hubble observed that more distant objects have a higher redshift in their light, which is evidence that our universe is expanding. Measuring the Doppler shift of distant stars has helped approximate the size of the universe, and it is very large indeed!

We said that the size of the universe is the largest thing we can measure, and it is so large that astronomers have to use a very big unit of distance called the *light year*. Physicists found that light travels at a constant speed of approximately 186,000 miles per second, or 300,000,000 meters per second in the metric system. Astronomers use the unit of the light-year when talking about huge distances. A light-year is the distance that light travels in one year. Let's find out how many miles are in a light-year:

<u>Example of a really big number</u>

<u>Problem</u>

Find the number of miles in a light-year

<u>Solution</u>

We know light travels at approximately 186,000 miles a second. To answer the problem of how many miles light will travel in a year, we must find the number of seconds in a year. There are 60 seconds in a minute, 60 minutes in an hour, 24 hours in a day, and 365 days in a year. To find the number of seconds in a year, multiply those numbers together. So we have 60 x 60 x 24 x 365 = 31,536,000 seconds in a year. Now that we know the number of seconds in a year, the final step is to use that number and multiply it by the speed of light per second. 31,536,000 seconds in a year times 186,000 miles per second = 5,865,696,000,000 miles in a light-year.

Here's an interesting four-dimensional fact: If a star one light year away stopped shining, it would take one year in time for us to notice because it is so many miles away. Even our own sun's light would continue to shine on us for 8 minutes after it was to shut off. Let's hope that doesn't happen because it would get very dark and cold. The picture in Figure 1.4 is a wide-angle view of the Crab Nebula taken from the Hubble telescope. The Crab Nebula is the result of a star exploding into a supernova.

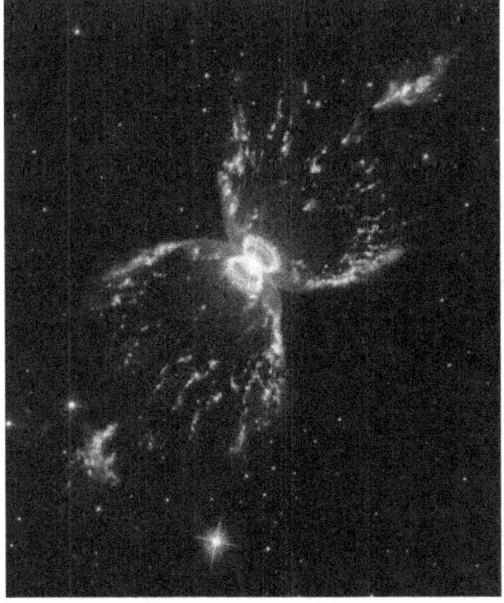

Figure 1.4. Crab Nebula

With the Doppler Effect's help, scientists have determined that the most distant observable objects in the universe are about 15 billion light-years away. When we look at them, we see the light emitted more than 15 billion years ago. This is another example of the fourth dimension, and we are able to see into the past. Even though the observable universe is about 15 billion light-years in all directions, and if it were a sphere, its size would be 30 billion light-years in diameter. Many recent theories say that it is much larger. (We are not at the center of the Universe, even though we act as though we are.) The generally accepted theory puts the diameter of the entire universe at almost 93 billion light-years. We will explore this idea more in the next chapter. Many of us have difficulty thinking about large numbers in the millions and billions and associating them with reality. In the U.S. Congress, they deal with massive government spending budgets and have an old saying, "A few billion here and a few billion there, pretty soon you're talking real money."

As numbers get huge, dealing with them becomes problematic. Scientists and mathematicians came up with a convenient way to express large and small numbers called *Scientific Notation.* The rule for scientific notation is that you would write one number, followed by the decimal point. To the right of the decimal point is where you would put the rest of the numbers, with a multiplier and exponent showing how many places the decimal point had been moved. In our above example, the number of miles in a light-year would be expressed in scientific notation as $5.865,696,000,000 \times 10^{12}$. Eliminating the commas and zeros would make it even more convenient, and 5.865696×10^{12} is a better way to show it.

Even though physicists have measured the speed of light with a great deal of accuracy, we used the common approximation for the speed of light in our problem and thus are limited. There is always likely to be some approximation and error in any number value. In calculating a problem where one of the numbers' accuracy is limited more than others, you must round off the answer to the least amount of accuracy. This limitation is called significant digits. Even though a calculator may seem to give a very accurate result, your most inaccurate number used in solving the problem limits the actual degree of accuracy. It is OK to use more than the allowed significant digits in intermediate calculations, but only the correct amount of significant digits should be given in the final result.

<u>Example of dealing with significant digits</u>

<u>Problem</u>

Find the diagonal distance of a TV screen. The length of the bottom is 83.527 cm and the side is 57.8 cm. Express the result both with and without Scientific Notation.

<u>Solution</u>

Since we are dealing with a plane and are given the X and Y values, just plug them into the Pythagorean Theorem. (Note: when there is no Z direction, use the formula shown below. It is the standard form.)

$$d = \sqrt{X^2 + Y^2}$$

$$d = \sqrt{83.527^2 + 57.8^2}$$

When we solve on the calculator the answer displayed is:

d = 101.5755863 cm

However, our accuracy is limited by 3 significant digits because the number 57.8 given in the problem has only 3 significant digits. The answer should be written as:

d = 102 cm.

This is the size of a 40-inch TV screen expressed in centimeters.

In Scientific Notation the result would be:

$$d = 1.02 \times 10^2 \text{ cm.}$$

Sections 1.3. Really small things

We have all experienced a cold winter day, but it's nothing to compare with the frigid temperatures in space. We need to understand a little about temperature from a physics perspective before we go on. Heat due to friction occurs at the atomic level. When you place your hand on an object at room temperature, the atoms and molecules are randomly vibrating. An object on fire would have the atoms vibrating vigorously as compared to objects at room temperature. The vibrations that cause heat always give energy to a system of lower energy. Something cold would have less vibrational friction, and it could absorb energy and heat up. From this discussion, we could surmise that no heat exists in space because there is no material to vibrate; however, there is no such thing as perfection in the universe!

There is no perfect vacuum, and outer space is not perfectly empty between celestial objects; there is a tiny amount of matter. Heat transfer doesn't only come from mechanical vibrations. The majority of the heat we receive from the sun is transferred to us in the form of electromagnetic waves. The electromagnetic energy is then transformed into mechanical energy. (Through reciprocity, mechanical energy could also produce electromagnetism energy.)

There are a few ways to quantify heat numerically. In the United States, we use the Fahrenheit scale to describe temperature, while the majority of the rest of the world uses the Celsius (centigrade) scale. There is another scale called Kelvin used mainly in science. Both Fahrenheit and Celsius have negative values, but Kelvin does not. The Kelvin scale starts at *absolute zero*. Absolute zero would be perfectly cold with absolutely no movement to produce friction. As we said before, nothing is perfect. An object might approach a reading of zero degrees Kelvin but never really get there. It could be argued that if absolute zero were to occur, objects would fall apart. When you touch an object, you are feeling mostly empty matter. Atoms are mostly empty space where electrons rotate so rapidly they form a cloud surrounding a central nucleus of protons and neutrons. If the motion were not there, the object would crumble. The temperature in interstellar space is not at, but very close to absolute zero. The latest experiments put the cold of space at about 3 degrees K. Water freezes at 0 degrees Celsius, which is 273 degrees Kelvin. So the temperature of interstellar space in Celsius is −270 degrees. In Fahrenheit, that is a reading of −454. The temperature is indeed very cold. It's far below zero on both the Fahrenheit and Celsius scales, but not truly a negative temperature. A negative temperature implies the opposite of temperature, and there is a temperature everywhere, even in space, albeit a very cold one.

Many people think that negative numbers are always small. Let's examine how a mathematician might explain a negative number. If we draw a number line showing a few integers in both the positive and negative direction, it looks as indicated below in Figure 1.5

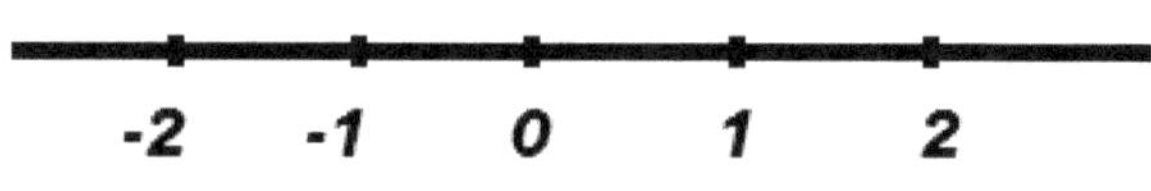

A section of the real number line

The real number line idea is that all numbers that exist are located somewhere on the real number line. If there are any numbers above or below the line, they are regarded as imaginary. We will discuss the imaginary numbers later. Each real number is associated with a point on the line, and each point is infinitely small. The line continues infinitely long in both the right and left direction. In the figure, integers are shown, but the line is made up of all possible real numbers. Fractions are located between the integers. Really small positive numbers are located on the number line between 0 and 1, and really small negative numbers are located between 0 and −1. Since each number is represented on the line by a dot that is infinitely small, the smallest numbers are the ones approaching zero from both the positive and negative directions.

Negative numbers are not necessarily small but are just in the opposite direction from the positive side of the number line. Each number on one side has an opposite on the other so that when they are combined, they will cancel out to become zero. In the example problem we looked at earlier of us walking around a room, if we start at the origin and take 3 steps to the right on the x-axis, and then take 3 steps in the negative direction, we are back at the origin. Another way to explain negative numbers is to reference Washington, as we did earlier in the chapter. The U.S. federal debt as of 2020 is over 27 trillion dollars. That means that the federal government owes that amount of money to creditors. That is a negative number, but it most certainly is not small!

<u>Example of a negative number shown in Scientific Notation</u>

<u>Problem</u>

Although it grows larger every day, according to the Congressional Budget Office, the current federal budget deficit before the latest round of stimulus for the fiscal year stood at 2.8 trillion dollars, express this number in scientific notation.

<u>Solution</u>

2.8 trillion can be written out as 2,800,000,000,000. When you write the number out the way we did, you can start at the decimal point and count the number of places to the left until we get to the scientific notation format. We want one number on the left of the decimal point, followed by the significant digits, and a multiplier and exponent showing how far the decimal point was moved.

In the above example, 2.8 trillion is equivalent to 2.8×10^{12} in Scientific Notation. To answer our problem correctly, however, we need to put a negative sign in front of the number because it is not money we have, it is money we owe. -2.8×10^{12} dollars is the answer to the problem.

If the number that you are trying to convert to Scientific Notation is not too large, it can be put into a calculator, and the calculator can make the conversion for you. Even when making conversions longhand, we should have a feel for the number prefixes like thousand, million, billion, and trillion. Although a little harder to remember how to write, we should also have a feel for the small number prefixes like thousandth, millionth, billionth, and trillionth.

In technology, to make working with large and small numbers more convenient, we use a similar process to Scientific Notation, called *Engineering Notation*. With this method, we work in groups of three number places so that our whole number is contained within a group of three places in one of the engineering prefixes. The decimal point is always located after a prefix, followed by the significant digits to the right of the decimal point. Engineering Notation is not all that dissimilar from Scientific Notation, except that we represent a nonfractional whole number in one prefix. The most common Engineering Notation prefixes are shown below.

<u>Engineering Notation</u>

For large numbers:

Trillion = Tera = $\times 10^{12}$

Billion = Giga = $\times 10^{9}$

Million = Mega = $\times 10^{6}$

Thousand = Kilo = $\times 10^{3}$

Units under a thousand don't have a prefix. The exponent is $\times 10^{0}$ for units.
The following are used for small numbers between 1 and zero:

Thousandth = Milli = $\times 10^{-3}$

Millionth = Micro = $\times 10^{-6}$

Billionth = Nano = $\times 10^{-9}$

Trillionth = Pico = $\times 10^{-12}$

Remember the number line? Positive fractional numbers are located between 1 and zero, and negative fractional numbers are between zero and negative 1.

Negative numbers would have a negative sign in front of the entire number. A small number of less than one unit would have a negative exponent. The negative exponent in math signifies a fractional number. In looking at our chart, an example would be the number 0.001 or 1 thousandth. In Engineering Notation, we would call it 1 milli, and it would be written as 1×10^{-3}. We moved the decimal place three places to the right to form a group of three, and the whole number is contained in the prefix. Since we moved the decimal point to the right, the exponent is negative, signifying a number of less than one unit.

The Scientific Notation rule about only having one digit to the left of the decimal point does not apply in Engineering Notation. In the engineering method, you stay in a group of three and work with just one prefix. It pretty much follows the way we deal with numbers in everyday life, except we use different prefix names and keep it in terms of only one prefix. We will use Engineering Notation throughout the remainder of this text.

<u>Example of using Engineering Notation for large and small numbers</u>

<u>Example of a large number</u>

If we wanted to talk about a number like 10,100,000 in normal language, we would say it was 10 million 100 thousand. In Engineering Notation, we would say the number is 10.1 Mega. Using Engineering Notation makes it easy for us to get a handle of dealing with exponents since we group them into groups of three. The number would be written as 10.1×10^{6}.

<u>Example of a small number</u>

In electronics, we sometimes use components called capacitors, which can pass AC but block DC. Some capacitors can be tiny in value. The unit of capacitance is the Farad, and we sometimes deal with values as small as 1 picoFarad. The number is 0.000,000,000,001 Farad. As expressed in Engineering Notation, it is the number 1 with an exponent of negative 12 after the multiplier. As you move the decimal point in the number to the right, you will count 12 places. Check back to the conversion chart if you are unsure about this.

Tip: when entering exponents on a calculator, do not enter x 10, instead enter the base number and press the exponent button. It may be labeled exp.

Chapter One summary.

For some people in the field of technology who may not be numerically oriented, we find that there is actually is a reason for learning about mathematics. We use the x, y, z coordinates in CNC machining and 3-D printing. We routinely encounter very large and very small numbers, and rounding off digits will lower the degree of accuracy so that we must keep the final significant digits in a problem limited to our least accurate input number. To aid in working with large and small numbers in engineering, we use a system similar to Scientific Notation called Engineering Notation.

Chapter 2

The need for speed

Section 2.1. The formal concept of speed

We can all give examples of speed, and you can get a speeding ticket if you exceed the posted speed limit in your car. The speed limits in the U.S. are posted as miles per hour. So it seems that speed is distance divided by time, $S = \dfrac{miles}{hour}$. A typical freeway speed is 65 miles/hour, or $S = \dfrac{65\,miles}{1\,hour}$.

<u>A typical math word problem about speed</u>

<u>Problem</u>

If you were driving a red car and leaving Walla Walla Washington at 10:32 AM, how many miles would you travel in 30 minutes if your speed was 60 mph?

<u>Solution</u>

There is a lot of irrelevant information in the scenario that we will ignore. The question we must answer is the distance traveled. We may be able to figure out the answer without using a formula, but let's look at how it would be done in a formal way. We know S = D/T, but we must rearrange the formula to solve for distance D = ST. Let's take a moment to look at how a mathematician might explain how the original formula is manipulated. In order to solve for D, it must be moved to one side of the equation where it will be alone. This process can be accomplished by multiplying both sides of the equation by T. This is allowed in math because as long as you do the same thing on both sides of the equation, it will remain in balance. So, we have the following algebraic manipulation of the original formula in order to solve for distance:

$$S = \frac{D}{T}$$

$$S(T) = \frac{D(T)}{T} \text{ and } \frac{T}{T} = 1$$

$$ST = D$$

$$D = ST$$

When solving problems, always make sure to use the proper units. You can't compare apples and oranges. Let's convert the time from minutes to hours like a mathematician with way too much time on his hands. We will convert it formally, and we will use a *proportion*. Many problems can be solved using proportions. They prove themselves to be a potent math tool. The goal of the challenge is to find what portion of an hour, 30 minutes would be. We will call it X.

$$\frac{X\,hours}{30\,min} = \frac{1\,hour}{60\,min}$$

In words, we say X hours is to 30 minutes, as 1 hour is to 60 minutes. We now go through the same procedure as we did in the above algebra manipulation, but this time we solve for X:

$$X\,hours = \frac{30}{60}\ hours$$

Reducing the fraction to lowest terms:

$$X\,hours = \frac{1}{2}, \text{ or } X\,hours = 0.5$$

Now that we have the units correct, we can solve our problem. Remember, we are trying to solve for the distance traveled in 30 minutes if we were moving at 60 miles per hour.

$$D = ST$$

$$D = (60)(0.5) = 30 \text{ miles}$$

With no hobbies to occupy their time, some mathematicians would take the solution a step farther and carry the units through the problem to verify the proper units. This process is a good way to solve a problem if you are unsure of what unit should be associated with the final answer.

<u>Example of carrying units through</u>

$$D = (60\ \frac{miles}{hours})\ (0.5\ hours)$$

The *hours* over *hours* cancel out.

$$D = 30\ miles$$

We have been using *quantitative* analysis to solve the problem, but let's take a moment to apply a *qualitative* analysis of the equation at the very beginning. In the formula we found from miles per hour, $S = \frac{D}{T}$, notice the S and T are *inverse* to each other because S is related to $\frac{1}{T}$.

<u>Qualitative analysis</u>

What if we wanted to travel a fixed distance of 30 miles and know how to get there in less time?

$$S\uparrow = \frac{30\,miles}{T\downarrow}$$

For the equation to remain in balance, as the time (T) tends to go down to become less, the speed (S) must go up.

That proves what we all know: our speed would need to increase to get somewhere in less time. In a sick sort of way, playing around with formulas gets to be fun. It's like any game; you need to understand the game's basic rules before you can play well. And the more that you practice at it, the better you will get. See the appendix of the book for more about math rules.

Section 2.2. Maximum speed in the universe

Now that we really understand and can precisely quantify the concept of speed, let's really break the 65-mph speed limit! Earlier, we said that the fastest thing in the universe was the speed of light in a vacuum. Physics experiments measured it to be approximately 186,000 miles/second. We also said that nothing in the universe is perfect, and even space is not a perfect vacuum. One could argue that the speed of light in a vacuum never really occurs in the universe. A real conundrum is as follows: If you were going nearly the speed of light in a rocket ship with headlights on, how fast would the light from the headlights be traveling? It seems reasonable to assume the speed would be almost twice the speed of light, roughly 372,000 miles/second. A conundrum is a riddle, and this is an especially good one. According to Albert Einstein, the answer to our problem is that it is all relative.

Remember when we were finding our way around a room in the last chapter, and we said we had to give our coordinates (X, Y, Z) in terms of a reference? We used the origin (0, 0, 0) as a starting point, and then we were able to quantify where we would end up after going a combination of distances in the X, Y, and Z directions. Without that initial reference, we wouldn't be able to explain our new position. The speed of light in a vacuum is absolute. There is a speed limit in the universe, and it is the speed of light in a vacuum unless you encounter something like a wormhole where things get really wacky. In our problem, we must consider the reference. There is no perfectly unique solution to this problem since the answer depends on the point of reference. Remember that we said that speed is distance divided by time $S = D/T$, as in miles per hour? A funny thing happens as objects with mass speed up, they get heavier, and time slows down. You may have heard that light is made up of particles called *photons*. Photons are very strange because they are a particle that has no mass. Light is also odd because it has a duality and behaves as though it is made up of both particles and waves. If you have a vivid imagination, you might think of the photons as surfers riding the waves at light speed! Einstein's theory of relativity can be demonstrated at speeds far below the speed of light. If you had precise timekeeping equipment onboard a super-high-speed aircraft, you could measure the slow-down in time with reference to someone on the ground. This may be the answer to the fountain of youth – *move very fast*! Time would appear normal in relation to you on the plane, but with reference to the observer on the ground, time would have slowed down for you. When you land, you would not have aged as much as the people on the ground. Don't put away the Botox yet; the difference would be so small that it would be difficult to measure. The overarching concept of Einstein's theory of relativity is that you must have a point of reference.

Now back to our conundrum, our riddle. If we keep in mind that with the theory of relativity, observations would take place in relation to a reference, then we can explain what happens in the rocket ship scenario. Because time is slow for them, the light from the headlights shines normally to the observers in the rocket ship. To observers on the ground, both the rocket ship and its lights would be traveling at nearly the speed of light. Einstein's theory also shows that as matter goes really fast, it gets really heavy. If matter were to go the absolute speed of light, its weight would be infinite. To move an infinite weight, you would need an infinite amount of energy. The chances are slim that you will ever get a ticket for breaking the speed limit of light in outer space!

We use many points of reference in our daily lives. As an example, you may have heard that the power companies in the United States provide 120 volts to our homes. That voltage coming from our wall sockets is referenced to *ground*. It means that there is a difference of potential of 120 volts between the hot wire and neutral or ground. This construct explains how birds can sit on electric wires that may not be insulated and not get shocked. There is no difference of potential across the birds. They have a different reference. But there would be a voltage between their location on the wire and the Earth.

Section 2.3. The speed limit of a computer

Sometimes a fast speed is necessary to arrive in a timely fashion when traveling on the roadways. We may want to go as fast as possible without breaking the speed limit. Speed is also vital in computer processing. Individually coded instructions execute as a computer program runs. Instructions are written in a language that mimics speech but with a syntax structure that can be interpreted by the computer processor. The instruction code must execute line-by-line as the program runs. These events happen behind the scenes every time you use a computer. The faster the code can be executed, the faster the program will run. Programmers should always try to write code to be as efficient as possible. To consider the fastest computing speed when dealing with a computer's electronic hardware, our previous discussion of the speed of light is helpful since we said that electricity moves through wires at roughly three-quarters of the speed of light. The actual signal speed through a wire is quite variable and depends on several factors. New materials are being developed in *photonics* research, which will use light flow through computer processors for the fastest speed possible. Photonics is used in fiber optic cables to produce blazing-fast Internet connections, but the light must be generated and passed through AND, OR, and NOT Logic gates in a computer. We will examine Logic Gates in later chapters. But for now, let's discuss how the electricity flows through the wiring of today's computers.

In the last chapter, we talked about the atom having a central core called the nucleus that contains protons and neutrons, surrounded by orbiting electrons. The relative numbers of these subatomic particles determine what element the atom will comprise. Protons have a positive electric force charge, neutrons have no charge, and electrons are negatively charged. Just as in romance, opposites attract. Opposite charges in the atom attract, so there is an attractive force between the protons and electrons. One of the earliest ways to describe how atoms are constructed was to compare them to the structure of the solar system. The protons and neutrons that comprise the nucleus compared to the sun, and the electrons were considered to be in orbits like the solar system's planets. In an atom, the electric force swamps the gravitation force. Thinking that the nucleus in an atom is like the sun at the center of the solar system is a good simplification. However, we now know that the electrons are not in 2-dimensional orbits like the planets, but instead envelop the nucleus in a spherical shape, sometimes called a shell. The word *cloud* is a far better representation because the electrons whiz around in a spherical shell that has some depth. The clouds may contain more than one electron at slightly different distances from the nucleus. As the maximum number of electrons per cloud is achieved, new clouds farther out from the nucleus form for the heavier elements. The clouds are discrete and do not mix together. They are said to form at different energy levels, with the higher energy levels being farther away in the distance from the nucleus.

The simplest atom comprises the element hydrogen. It has only one proton in its nucleus, surrounded by one electron whizzing around at the lowest electron energy level. Hydrogen is the essential element in the universe and is the most common fuel for stars like our sun. Hydrogen is also being used in fuel cell technology as a form of alternative energy to power over-the-road vehicles. Electricity is needed in a fuel cell to separate the hydrogen and oxygen in water (H_2O) in order to make combustible gases. The electric process of separating the gasses from the water is called *electrolysis*. Care must be taken when working with fuel cells because the gasses are incredibly flammable. As a star burns hydrogen at extremely high temperatures in a nuclear reaction, a process called *fusion* takes place that produces helium. Helium is the next heavier element and is made from two protons in the nucleus and two electrons in the lowest energy cloud. As stars burn the heavier elements through the fusion process, even more massive elements are produced. The heaviest naturally occurring element is uranium, with 92 protons and 92 electrons. Using particle accelerators called *cyclotrons*, commonly referred to as atom-smashers, scientists claim to have observed non-naturally occurring elements with as many as 118 protons and complementary electrons. Notice we have equal numbers of protons and electrons because for an atom to be stable, the positive and negative charges must be equal and opposite. If a material does not have equal charges, it is said to be an ion. Ionic material will have an overall positive charge if it has missing electrons and a negative charge if there are extra electrons.

Since there is an electric force of attraction between protons and electrons, the question might be asked as to why the electrons don't move toward to combine with the protons in the atom's nucleus? Another good question is: why don't the planets leave their orbits and fly into the sun because of the pull of gravity? Now is a good time to introduce the concept of a vector. Earlier, we examined speed. Speed is called a scalar quantity. In an example problem that we worked earlier, we left Walla Walla, Washington, at a speed of 60 mph, and speed is a *scalar* quantity. If we were given a direction in the problem, then it would be called a *vector*. A vector has both magnitude and direction. An example of this appeared in an old movie titled *Airplane.* In the movie, the pilot asks the navigator, whose name is Victor, "what's the vector, Victor?" That's a great joke if you're a nerdy physicist. Let's now apply vector analysis in Figure 2.1 to see why electrons aren't pulled into the nucleus of atoms, and planets into the sun.

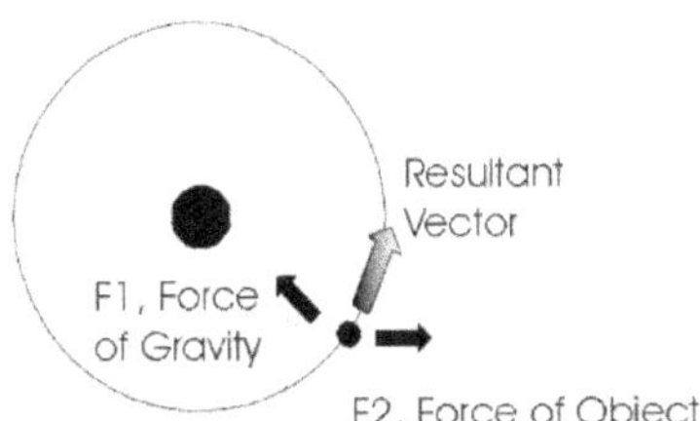

Figure 2.1. Hydrogen atom

When something is in a stable orbit around an object, two vector forces come in to play. They are forces of acceleration, and they are vector forces because they have magnitude and direction. In the above figure, F1 is the acceleration due to gravity. It is trying to make the orbiting object fall. Vector force F2 is the acceleration of the object trying to send the orbiting object in a straight line. We say that these forces are due to acceleration because the *vector* directions are continually changing. The magnitudes stay constant and are equal. This situation causes a resultant vector that maintains the orbital rotation as F1 and F2 shift directions. An orbiting object may be the closest thing to a perpetual motion machine. However, we must keep in mind that this is a closed system with almost no loss of energy. In a machine, the moving parts always cause friction that results in energy loss radiated as heat.

Electrons are in orbit around the nucleus of the atoms in current-carrying conductors in a computer. Electrons are small when compared to protons. They are 1800 times less massive than protons and are the current carriers. Copper is the most popular conductor used in wiring. It has 29 protons with 29 electrons forming 4 discrete spherical clouds around the nucleus. The cloud at the lowest energy level nearest the nucleus contains 2 electrons. In the next higher cloud, there are 8, followed by 18. Finally, the single electron we are interested in is located in the outer cloud called the *valance* cloud. It is the cloud most distant from the nucleus with the highest energy level. For the element copper, any forces outside of the atom can easily move the electron in the valence cloud out of the atom. The electron positioned there has a high energy level, and there is not as much of an attractive force from the protons in the nucleus. Another reason is that elements in the universe are very stable if they have 8 electrons in the valence cloud. These elements have the electrical characteristic of being an *insulator*. When there is only one electron in the outer cloud, it is not very stable, and the electron can be dislodged rather easily. When energy is applied to a conductor, the nearest atom's valence electron becomes dislodged and freed from the atom. It will then fall into the next closest atom, thereby dislodging that atom's valence electron, and on and on we go. Electrons entering the conductor equal those leaving, so the conductor stays in balance.

Figure 2.2. Satellite dish

Most long-distance telephone calls travel the majority of the path via satellite or fiber optic cable rather than by wire. Speed is not a consideration for telephone calls, but high data speed gives increased channel capacity. More communication channels are available in this way. There is a slight time lag, even at light speed. Communication satellites send signals at the speed of light, but they are very far away. They are placed in orbit approximately 20,000 miles above the equator so that they can stay in the same relative position in the sky as the Earth rotates. This effect is called a *geosynchronous* orbit. Even at light speed, to cover the 40,000-mile round trip takes about 0.2 seconds. Additional time is also needed for the signals to transverse electronic circuitry on both uplinks and downlinks. If you see a foreign correspondent report on the six o'clock News taking a little time to answer a question from the TV anchor, it may be due to this process.

The reason that electricity moves so fast is that the electrons don't really move very far as they flow through a wire. If you were to place a telephone call across the world on a wire, it wouldn't take an hour for their phone to ring. There would be a slightly longer delay than with direct satellite communication, but as we said earlier, electric signals move through wires at roughly three-quarters the speed of light. There is a novelty item that works mechanically in somewhat the same way as electrons flow electrically through a wire. It is a series of six or so balls suspended by strings. When you raise one and let it go, it hits the stationary balls at the bottom, causing the ball on the far end to instantly pop into the air. With electricity, when you turn on a light in your house, the electrons that produce the light in the bulb don't travel all the way from the power generation plant to your home; they just knock into the closest atom at the power plant, and the closest electron to the bulb pops out of the wire.

Earlier, we said the particles in a stable orbit are in a closed system and don't lose energy. As electrons flow through wiring, there is some friction to overcome, and energy is lost as heat. The friction is due to the random vibrations that occur inside the wire due to the wire being at room temperature. As the wire absorbs heat energy from the ambient air, the material's random motion vibrations increase at the atomic level. The friction in electric circuits is called *resistance*. By cooling the wire, it is possible to decrease resistance. Superconductors use this concept and special conductive materials to achieve near-perfect conductance. The problem with cooling wires is that it is impractical for most circumstances. Computers contain fans for cooling, but this to keep the CPU and other electronic components from overheating. Although having better conduction in the wiring of a computer would help with energy efficiency and save power, the speed of the current in a wire is limited by the layout of the busses. With a fast processor, busses are the bottleneck in a computer's processing of the signals due to electric capacitive and inductive interaction limiting speed because of crosstalk.

34

Chapter Two summary

We have all traveled in cars that hopefully were going under the speed limit. Word problems can sometimes be challenging, but the process can be simplified with work done slowly and methodically. The first step is to understand the problem, develop a solution by assigning variables to the known quantities, and then solve the unknown through algebraic manipulation. The choice of letters you use are arbitrary, but they should reflect the names of the quantities. Our example problem at the beginning of the chapter would have been more standardized if we used lower-case letters because they more represent instantaneous and changing quantities. Solving for a numerical result is a *quantitative* analysis, whereas describing the nature of something is called *qualitative* analysis. We usually need a point of reference in both cases. Electronic signals' speed is extremely fast, but in circumstances where wiring is tightly packed together, electronic properties of capacitive and inductive interaction will limit the speed and cause a slowdown.

Chapter 3
Parts of a computer

Section 3.1. Computer background

Computers have a clock, and the faster it can tick, the faster the machine can operate. We're not talking about the computer clock that gives you the time; we're talking about a behind the scenes signal that is like a metronome used in music. A metronome keeps the beat in music; all of the music is referenced to it. The hardware clock sends out a high speed on/off signal called a square wave in a computer, as shown in Figure 3.1. The signal is on for 50% of the time and off for 50% of the time. We could also say that it has a 50% *duty cycle*. One cycle of on-and-off time per second is called one *Hertz*. Microcontroller clocks could run as slow as the low MegaHertz range, while the clocks used in computers operate in the low GigaHertz range.

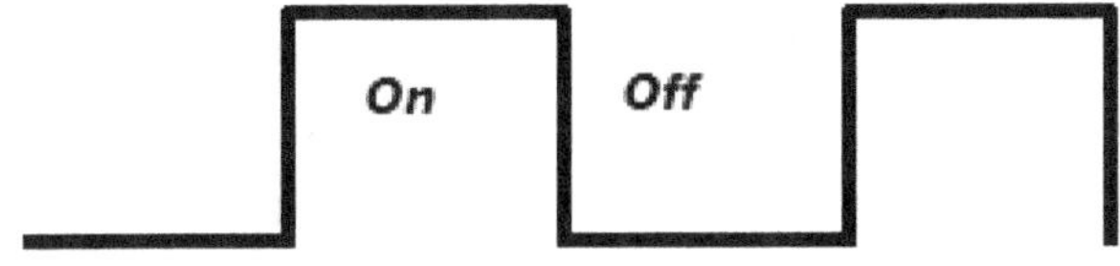

Figure 3.1 Square wave

You may have heard that computer language is made up of ones and zeros. Many programming languages closely mimic English but have a specific syntax so that there is no ambiguity in the meaning of the words. The programming languages are ultimately changed into ones and zeros by a special program called a compiler. The computer hardware uses ones and zeros because of the simplicity of the electronic circuits. Think of a simple light switch on the wall. A data one (sometimes called a high) is when the circuit is on, and a data zero (sometimes called a low) is when the circuit is off. Given our progress in the field of electronics, this may not be the most elegant system. The use of only two logic levels is a throwback to the early days of computing when monstrous machines crunched numbers with vacuum tubes and mechanical relays. As we make strides towards quantum computing, the fundamental data are called qubits and can have three states.

Figure 3.2. ENIAC Computer

In the U.S. Army photo in Figure 3.2, one part of the first electronic computer is shown. The army used the ENIAC computer just after World War II for projectile motion calculations. Today, an inexpensive calculator is by far a superior computing device. The ones and zeros system is called *binary* because there are two possibilities. The system is easy for electronic circuits to process and for numbers to be stored in memory. In the days of the early computers, light was the medium to hold memory. It was not fiber optics, but instead holes, or a lack of them, on a punched card or reel of paper tape, which acted as the permanent memory. A light was on one side of the paper, and on the other was a light pickup sensor used to detect each of the two states.

After that system, magnetic storage was developed and still is used today in mechanical hard drives. In magnetic storage, an aluminum platter is coated with fine metallic particles. A small coil of wire called a *head* uses the property of electromagnetism to write and read data. Writing is the process of saving the data to the platter. Reading is the process of opening a data file, thus detecting the ones and zeros from the spinning platter. In writing, the electric current's polarity through the wire creates a magnetic force to cause the magnetic pole orientation of an area in one direction on the metallic surface to be considered a one, and in the other direction, a zero. When reading, the magnetized moving areas on the platter causes the appropriate polarity of current to be induced into the wiring of the head. This type of vice versa action is called *reciprocity*. Another example of reciprocity is a radio antenna; it can work for both transmitting and receiving information.

A computer hard drive head is located on a lever that can pivot along the radius of the platter so that it can reach any point as the platter rotates. For the electromagnetic interaction to occur, the head must be extremely close to the platter's surface. For fast data access, the platters in modern hard drives continually rotate at very high speeds of 5400, 7200, or 10,000 rpm. There can be no contact with the head when the platter is rotating at full-speed to limit friction. There is a landing zone area on the platter for when the head is parked. When it begins spinning fast enough, the Bernoulli Effect lifts the head, just as an airplane wing lifts an airplane into the sky. The tolerance is so tight; however, the drives are assembled in a clean-room environment, and if they are ever taken apart in the field, a speck of dust in the air could come between the rotating platter and the head, causing the drive to be damaged. As the cost decreases and storage capacity increases for solid-state hard drives, they will eventually replace mechanical hard drives. The solid state hard drive is similar to Flash technology used in removable flash drives, called thumb or jump drives. NAND gates, which we will study later, are used in solid state drives (SSD). A floating gate on a field-effect transistor either allows, or stops conduction, for the representation of the binary data. The main advantages of SSDs are speed, greater reliability due to no moving parts, and lower power consumption than mechanical hard disk drives (HDD).

In the last chapter, we described how the signals move through computer wiring. The wiring in a computer is called a *bus*. There are separate busses for power, data, control, and clock signals. Even though we found that electricity can move at nearly the speed of light in power line wires, busses in a computer are limited in their speeds due to the transitions of highs and lows and interactions between the wires, which we mentioned earlier. Many newer computers can have clock speeds in excess of 3 GigaHertz (GHz). Looking at the chart in chapter one, that's 3 billion clock cycles per second! You may have noticed the computer operates slowest when you are reading or writing to the mechanical hard drive in working with computers. Electricity and magnetism are very fast, but the limiting factor for a hard drive is that it is mechanical. Even with the platter rotating at a high speed of 10,000 rpm, some time is involved for the data area to rotate around to the head. The head also takes a little time to pivot along the radius of the platter. Periodically running the disc defragmenter program will help to keep the bits in the same general area.

The working memory called Random Access Memory (RAM) can operate much faster than a hard drive, especially a mechanical one. RAM has the drawback of being *volatile,* however, which means that the data is lost when the power is off. There are two types of RAM working memory. The RAM that is modular and usually can be upgraded in a PC is called DRAM (or SDRAM). DRAM stands for Dynamic Random Access Memory, and the S letter just means that it operates synchronously. DRAM has a few disadvantages because the transistors are field-effect like the SSDs. Field effect transistors (FET) are not the fastest switching transistors. Also, because DRAM uses an electronic component called a capacitor to store each bit of date, they periodically must be refreshed by rewriting the data every 64 milliseconds for DDR2 modules. The density is high, and the cost is low for DRAM memory, making it the main working memory. The fastest working memory in a computer is SRAM, which stands for Static Random Access Memory. The transistors can use bipolar technology where a base current controls the conduction rather than the effect of a voltage field like in the FET. SRAM is used in Cache memory, which is closely associated with the central processing unit (CPU). Cache memory is put at distinct distances from the processor. The level 1 (L1) cache is located inside of the processor IC, L2 cache could be on the chip or nearby on the motherboard. Sometimes there is also an L3 cache supply on the board.

Section 3.2. Parts of a personal computer

We covered a few of the parts of a computer. We discussed the wiring called busses, the hard drive used for long-term memory, and the short term but fast Random Access Memory (RAM). The RAM is sometimes also called the working memory because it is the section of the computer where the coded instructions and data reside as processing occurs. The instructions and data are pulled from the long term storage medium and stored in RAM to run. A computer would not even be able to boot up without the RAM. The execution of instructions and manipulation of data occurs in the Central Processing Unit (CPU). The CPU is located on the motherboard along with the *chipset* that helps the processor perform many auxiliary tasks. The motherboard is the main circuit board where everything connects. Most computers also have sound and video processing circuits on the motherboard. Computers with higher quality video and PCs used for gaming may have a separate circuit card to display better video graphics. The motherboard may also have sockets that can be used by expansion cards for a variety of different purposes. Very old PCs had ISA slots, then PCs were upgraded to the PCI standard slots for expansion cards. Now we use the very fast PCI Express bus (PCIe). The PCI bus was of a parallel design that periodically switched between devices, whereas PCIe is based on serial connections made by a host controller. Most PCs will have a DVD/CD ROM drive that also plugs into the motherboard. The power supply will connect to the motherboard and other equipment like the DVD/CD, hard drive, and any other internal modular devices. The power supply also supplies power to any USB devices that are connected. The power supply takes the 120 volts AC from the wall outlet and converts it to a constant $\pm$ 12, $\pm$ 5, and 3.3 volts DC. There may be voltage regulators located on the motherboard to convert to other voltages to power specific devices.

Power supplies with voltage regulation are used in almost every electronic appliance. The main purpose of a power supply is to convert from AC to DC. AC stands for Alternating Current and is used for power distribution by the electric company, and DC stands for Direct Current, which is used with most electronic components. AC switches direction almost like the square waves we discussed in the last section, except that square waves change between a higher and lower (or zero) voltage and do not change in polarity, but AC actually reverses direction and alternates between a positive and a negative voltage. Voltage is a difference of electrical potential that can produce a current through a device, hence the term alternating current. It seems that the voltages and currents would cancel out in this case. Early in chapter one, when we were discussing the number line with positive and negative numbers, we had an example of you walking in a room in the positive x-direction and then turning 180 degrees and walking the same distance in the opposite direction. You would have ended up at the same place where you first started, but you would have expended energy in the process. The process also takes some amount of time, and the vectors are not in opposite directions identically at the same time.

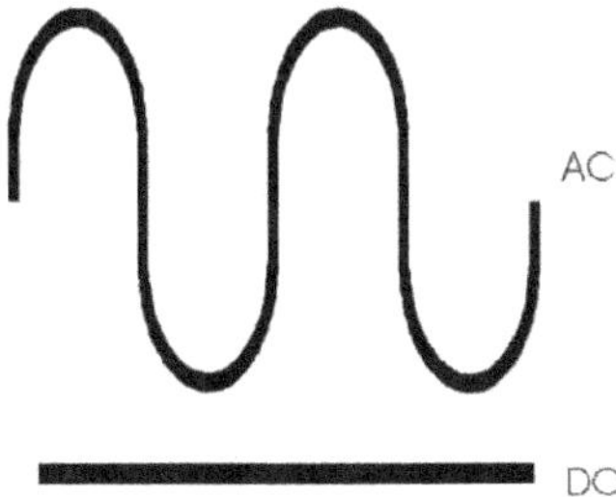

Figure 3.3. Sine wave and a DC voltage level

In figure 3.3, we view the two types of voltages as changes in their potential difference on the y axis with reference to time along the x axis. The DC is a direct voltage and current and keeps the same difference of potential level unless it is switched on and off as we did in a square wave. On the other hand, AC is quite complicated. Our drawing may seem like one positive and negative section makes a circular rotation spread out over time. This is almost true, but due to power generation equipment's nature, it is slightly elongated in height and is called a *sinusoid*. Like a circle, the sine has 360 degrees, and one rotation spread out over time is called a cycle, which has a specific time period. As with the square wave, the frequency has the unit of Hertz, and one Hertz is defined as one revolution or cycle per second. The frequency of electricity to our homes in the U.S. is 60 Hertz. It is 50 Hz in some other parts of the world. The voltage of the sine waves coming to U.S. homes actually reaches peaks at about 170 volts. When we say the voltage at a wall socket is 120 volts, we are talking about the voltage at the 70% value of the peak above zero. By default, we describe AC voltage in that way. The 70% measuring point is defined mathematically as the Root Means Square (RMS) of AC and has an equal heating effect to the same voltage value in DC.

AC is much more complicated than DC, and there was some debate before it was adopted as the standard way to distribute power throughout the world. In many other countries, they decided on different standards for the voltage and frequency, but all use AC even though we must have power supplies in electronic products to make a conversion from the AC to DC. The sensibility of using AC might be questioned, but there are a few underlining reasons why it was chosen over DC for transmission over distances. When we power a device, we could use a battery. Batteries provide DC electricity through chemical action in the cells. Once the chemical reaction occurs in a primary cell, the battery goes dead. In a secondary cell, connecting it to an external energy source and recharging the cell can rejuvenate the chemical action. In our power plants, the generation is accomplished using a device similar to a motor. Motors are yet another example of reciprocity because when they are connected to an electrical source, they spin; reciprocally, if you use a force to spin them, they will generate electricity. As with batteries, we are converting energy from one form to another. In the case of a generator, it is turning mechanical energy into electrical energy. There are many ways in nature to get a force to spin a wheel to generate electricity continuously. Some common examples are water flow in hydroelectric plants, or when steam is produced to spin a turban from burning coal, gas, or nuclear fuel, as well as the generation of power from wind energy. We use AC for power distribution because it can easily be converted to extremely high voltage to reduce power loss in the transmission lines and then brought back down to a reasonable voltage level to enter our homes. We mentioned earlier that wires have resistance due to the random vibration of the atoms from heat. This resistance in the wire causes some of the energy to be used in overcoming this opposition to electric current. The same amount of power can be delivered with lower losses through high voltage transmission lines because the current is minimized in the process. Even if we were to use DC for power distribution, we probably couldn't get away with not having power supplies in electronic devices because they perform the function of regulation of the voltages as well as conversion. As the conditions change at both the generating and receiving end (called the load), the voltages could fluctuate, and this is why regulation is needed. In a computer, it is especially true that the voltages levels must be tightly regulated since the ones and zeros are processed as voltage levels. However, some noise immunity is built into computer chip design.

An exciting trend is taking place as Internet connections are becoming faster. Expensive hardware and software will be able to be accessed externally from your machine. This process is called distributed or cloud computing. When we studied the make-up of atoms, we found that the electrons were in orbit in clouds around the nucleus, and there was an interaction of forces that formed the particles into a cohesive unit. In somewhat the same way, you can interact at a distance with different computer devices and not be too concerned about their actual physical location. This may make notebooks, laptops, and PCs less expensive. Their operation may become comparable to a high-end local computer since robust high-speed functions will be accessible. There may no longer be a need for a large hard drive, and even the speed of the processor in a local machine might not be as vital as it once was. We assume that your Internet connection is very high-speed since it will be essentially another type of computer bus.

Section 3.3. The cellphone, today's Swiss Army knife

The Swiss Army knife is a tiny gadget that holds a wide variety of small tools like a can opener, corkscrew, and other helpful devices. People are so attached to their cellphones that it almost seems like an extra appendage. The comparison to the Swiss Army knife is appropriate because the phone apps are small tools for simplifying our lives. The telephone function seems almost secondary, with texting being the preferred mode of communication. The cellphone is essentially a tiny computer containing a camera, music player, voice recorder, GPS, game device, etc.

The major part of a cellphone is a radio receiver/transmitter, but the phone is very much like a miniature PC connected to the Internet. With 5G technology, the connection speed is blazingly fast. The fantastic thing is that all the functions have been incorporated into one device. Some people feel that this kind of equipment is bound to change computing's nature since you will now no longer be physically bound to a computer. You may have binged watched rerun episodes of a TV show called *Star Trek* starring William Shatner during the pandemic lockdown. In that TV show and subsequent movies, futuristic members of a group of planets called *Star Fleet* explored the galaxy. They used a handheld device called a *communicator* for voice contacts, and a device called a *tricorder* for data. At the time of the early TV episodes, the only actual portable telephones were as large as a suitcase and located in the trunks of cars. The airtime was so expensive that very few people even knew the technology existed. In that era, data was only processed in computers owned by the government, schools, and large corporations. What was fanciful science fiction thirty plus years ago has become science fact! Traveling around the galaxy is still a problem because of the great distances that we spoke of earlier. No technology exists yet for physical matter to exceed the speed of light. But exchanging voice and data at light speed over small hand-held devices is so popular that almost everyone today is utilizing what was once only a dream of science fiction.

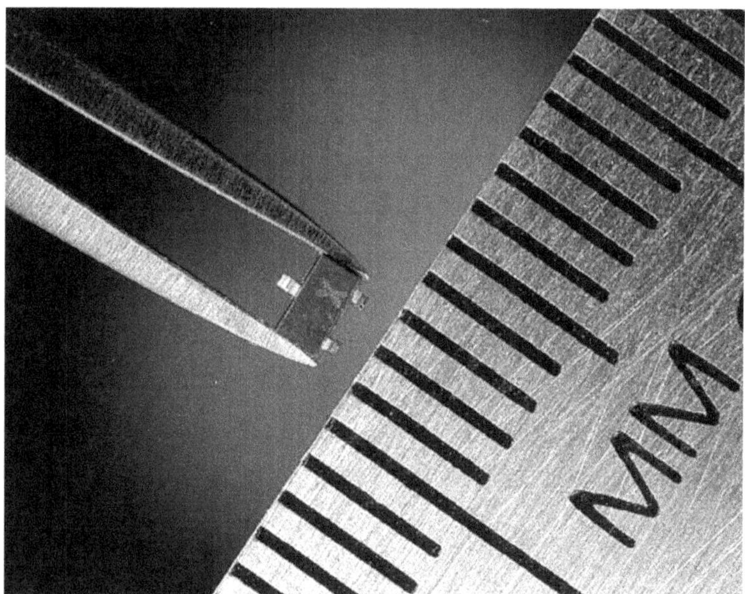

Figure 3.4. Surface mount transistor

The miniaturization of electronics, like the surface mount transistor pictured in Figure 3.4, and integrated circuits, first produced in 1958, provide the average consumer with robust capabilities from relatively inexpensive devices. Initially, the cost of many electronic products was high, but parts have gotten far less expensive. With worldwide acceptance and high demand, production costs have gone down with efficiencies introduced in manufacturing the devices in large volumes. Cellphones today might almost be considered a *commodity* item. A commodity is a part of everyday life and not thought of as anything special. But a lot is happening inside of a cellphone. Just taking a moment to look at voice communication technology, we have a radio receiver and transmitter.

Radio signals travel as waves in much the same way as light. They are below visible light in the electromagnetic spectrum and are all around us. We are not directly sensitive to them in the way that our eyes are sensitive to light. Our ears pick up low-frequency vibrations of the air that we interpret as sound, and our eyes see the light waves on the electromagnetic spectrum. But we need a radio receiver to pick up and gain information from radio waves. The waves that we are speaking of are sinusoidal like the AC power waves we discussed earlier. Unlike power or sinusoidal audio waves, electromagnetic waves do not need a conductor and can move freely through space at light speed. The electromagnetic waves are transverse. They have an electrical and a magnetic component occurring perpendicular to each other and to the direction of travel.

Transverse wave

For the cellphone to send and receive radio waves, it must have an antenna. An antenna is a transducer, a device that changes energy from one form to another. A simple example of how a transducer works is the tuning fork. When you strike a tuning fork, you are using mechanical energy to cause air vibration, which we hear as sound as the ears diagram vibrates sympathetically. The physical size of the tuning fork determines its tone. If the tuning fork is large, it will produce a low-frequency sound. Low frequencies have a long wavelength, with the waves spaced far apart. A small tuning fork delivers a higher frequency tone with a shorter wavelength. A radio antenna is a transducer that takes electric current flowing back-and-forth through it and converts it into electromagnetic energy that can propagate through space. Just as the tuning fork vibrates mechanically, the radio antenna seems to be vibrating electrically. And like the tuning fork, the waves' frequency is related to the antenna's physical size. The electromagnetic waves used for many cellphones are roughly at 1 GigaHertz, shown in the chart in chapter one to be one billion cycles per second. We can solve for the wavelength with the following formula:

$$\lambda = \frac{c}{f}$$, where λ (lambda) is the wavelength, and c is the speed of light.

<u>Problem</u>

Find the wavelength of a cellphone wave in inches.

<u>Solution</u>

     c = (186,000 miles per second)(5280 feet in a mile)(12 inches in a foot)

     λ = 11.78496 x 10^9 / 1 x 10^9 (we are using engineering notation.)

     λ = 11.8 inches

Often, the physical size of an antenna is restricted to either a half or quarter wavelength to save space and work at only a slight decrease in efficiency. Dividing 11.8 inches by 4 gives a size of slightly less than 3 inches for the size of a quarter wavelength antenna inside of your cell phone. Cellphones are not exactly at 1 GHz, but are tunable a little above and below that frequency. You can tune a radio, but you can't tune a fish. (Sorry for the bad joke.)

Your cellphone needs its own radio channel for you to have a private conversation. There was once a party-line service on the old hard-wired twisted-pair telephone system, sometimes called POTS for Plain Old Telephone Service. It was often the only option in rural areas and was offered as a low-cost alternative in some other locations, mainly until the 1960s. A party-line had users share a circuit. There was a selective ring so that a specific user knew that an incoming call was for them. To place an outgoing call, you would have to pick up the phone and make sure nobody else was using the line.

Even though cellphones occupy a narrow bandwidth, there is a limit to the number of available channels. However, the system allows for channels to be reused in geographically separate areas because of the low power used in transmissions. When you have a conversation, your phone connects to the nearest cell tower. There could be another conversation on your channel in a different city, but you will not interfere with each other. The system uses a beehive honeycomb pattern of cells, and the frequencies are coordinated, so adjoining cells use different channel frequencies to reduce crossover. When not in use, the cellphone is tuned to the primary call channel. When placing or receiving a call, the system assigns the user to an unoccupied talk channel. This process is sometimes noticeable as a slight connection delay when placing a call, or as interference on any nearby audio amplifier, a few seconds before the cellphone begins to ring. Cellphones also produce periodic interference with audio systems because the phone checks for a stronger cell site from time-to-time. As the signal level improves with another cell site location, your service will be *handed-off* and transferred to that site.

Cellphones initially used a time-sharing process called Time Domain Multiple Access (TDMA). With TDMA, the individual phone communication bits were transferred at predetermined time slots. Other individuals would have their bits interspersed during the other time slots. This process of time-sharing gives what seems like multiple channels on one frequency. Not every communication device needs a specific frequency. There is a communications system called *spread spectrum* where signals hop around or sweep back-and-forth. The transmitter and receiver must operate in synchronization and move to the changing frequencies simultaneously. The process may seem to be somewhat random and happens very quickly. These types of communications systems are nearly impossible to eavesdrop on and were initially designed for military and other high-security uses.

The next advancement after TDMA was to use Code Division Multiple Access (CDMA). This technology assigns a code to each individual phone and utilizes spread-spectrum channels. This technology is considered 3G, and 4G made advances in speed, but some providers pushed for LTE to replace 4G. The letter G stands for generation. LTE is an abbreviation for Long Term Evolution. It is based on an Internet Protocol (IP) structure and offers speed enhancements through digital signal processing. 5G began rolling out in 2018 and uses more bands of frequencies, including a new band starting at 24 GigaHertz, on the upper end of the cellphone frequency range. At super-high frequencies, the radio signals begin to take on some of the characteristics of light waves and don't pass as well through objects as do radio waves at lower frequencies. This band uses extremely localized communication cells by increasing the directivity of the link between the user and a cell tower. The power of a radio signal increases with directionality. Also, as frequency increases, the transmitted energy becomes higher, and there is some concern over health-related issues from the potent directional electromagnetic radiation needed in urban areas. 5G produces much faster throughput by utilizing a very wide bandwidth in the higher frequency ranges. 5G at these frequencies not only provides cellphone transmissions but also high-speed links for many Internet of Things (IoT) and computing devices. Under ideal conditions, it can offer data speeds up to gigabit throughput.

Chapter Three summary

Computers process instructions and manipulate data through busses connecting the different sections. The CPU can operate in the low GigaHertz range, but the busses and memory are slightly slower. The CPU works with helper chips on the motherboard called the chipset. Data is composed of ones and zeros and can be stored as magnetic areas called domains on mechanical hard drives, and as current flow through field-effect transistors in solid-state hard drives utilizing floating gates. Hard drives are for long term storage and nonvolatile. Working memory is volatile and comprised of DRAM FET type transistors working with a capacitor requiring continual refreshing, and as SRAM latching transistor circuits in different levels of cache memory to store ones and zeros as data bits.

The cellphone is a digital radio receiver/transmitter, receiving and sending logic levels similar to what we discussed in computer technology. The digital signals must be impressed on an analog radio carrier wave to propagate through space. Digital communication can be carried directly only if there is a hard-wired circuit or through a fiber optic cable; otherwise, it must be converted to an analog channel. The analog carrier is usually a simple high-frequency oscillation of AC current sent to an antenna. The digital logic that conveys information is the subject of the next chapter.

Chapter 4

Logic isn't just useful for computers

Section 4.1. Logic has been around a long time

The study of logic dates back to the days of ancient Greece, when many extremely complex problems were solved through deductive reasoning. In an era before the technology existed to enable elaborate experimentation, the application of analytical problem-solving methods produced amazing discoveries in science and mathematics. In today's technological world, engineers and technicians must always thoroughly think about the problem before settling on a solution. Understanding the problem is especially true when *troubleshooting* complex electronic circuits. A technician repairing a computer circuit board, for example, should identify the problem from the symptoms of the malfunction, and then use the schematic diagram of the circuit board to theoretically troubleshoot the problem before attempting any repairs. To do otherwise would entail swapping parts in-and-out, or making many unnecessary tests and measurements, both of which are inefficient strategies.

Logic is a pretty familiar term to us all, and sometimes it seems to be the same thing as common sense. Before we investigate computer logic, we should precisely understand what logic is all about. In a philosophical sense, logic is the process of applying an analysis using deductive reasoning. In the analytical thought process, making a deduction is not the same thing as making an assumption. In deductive reasoning, the background facts are clearly understood, and it is then possible to *extrapolate* the unknown.

Logical reasoning could be incorporated into an electronic device. In some Blu-Ray HD players, standard lower resolution DVDs are up-converted to HD quality levels. The higher resolution of HD is because the *pixel* spacing is very close together. A pixel is a contraction of words that means picture-element. It is the smallest unit that would be observable to you on a TV or computer monitor screen. If the screen was HD capable and could illuminate pixels that were very close together, the standard quality DVD could not produce that high of a resolution. Still, the missing pixel information can be *interpolated*. Interpolation is the process of examining the information about what is known and *extrapolating* new information. In essence, we will decide the unknown from the material that we know. If you want to argue that this is an assumption, consider it an assumption based on a good understanding of the known information.

Using the Standard DVD to HD TV screen scenario, let's look at ways to interpolate a missing pixel's brightness. If there were a few pixels to the left and right of a missing pixel, we could reason what our missing brightness might be by analyzing what we know about the brightness of the nearby pixels. A good mathematical approach might be to assign numerical values to different levels of intensity. Taking the pixel's value on the left and adding it to the value of the pixel on the right and then dividing the sum by 2 will give the average brightness, and we could use that value for the missing pixel.

$$\text{Average Brightness} = \frac{Pixel_{left} + Pixel_{right}}{2}$$

The process just outlined is an example of deductive reasoning and the use of an *algorithm*. In computer processing, algorithms are mathematical formulas or problem-solving methods that produce a result. Our algorithm's complexity could increase to consider more factors about the surrounding pixels to achieve a more accurate result. A computer chip called a microprocessor is often incorporated into the hardware to make these types of judgments possible in electronic devices, and a software program then controls the operation. Every time you push a button on a modern TV or use a remote control, your actions are accomplished through a microprocessor's reaction. They have become very commonplace, and some appliances such as stoves and refrigerators have tiny self-contained microprocessors termed microcontrollers. There may be more than a dozen microcontrollers in your car.

The logic of digital electronics is pretty straightforward, but the study of philosophical logic can be tricky because of the ambiguity of language. Miscommunication can lead to flaws in logical arguments called fallacies. In technology, we rely on mathematic principles to eliminate any confusion over words, and as we will see next, digital logic in electronics is very straightforward.

Section 4.2. Digital logic gates.

The mathematics of digital logic requires only a few functions that are described by operators. We use operators similar to what is used in arithmetic but with slightly different functions and names. An example of an arithmetic operator is the + sign. It tells us to add numbers. In digital logic, it means OR, and in binary arithmetic, we will learn later that 1 + 1 = 1. For now, let's look at the seven digital logic functions and their operators. We will also learn about their symbols and truth tables.

Logic Function	Operator	Example
NOT	Negation symbol (-)	NOT a = $\overline{a}$
OR	+	a OR b = a+b
AND	x	a AND b = ab

By combining the NOT with each of the other two logic functions, we have:

Logic Function	Meaning	Example
NOR	NOT OR	$\overline{a+b}$
NAND	NOT AND	$\overline{ab}$

There are only two other logic functions, the exclusive OR and the exclusive NOR:

Logic Function	Meaning	Example
XOR	a OR b, but not both	$a \oplus b$
XNOR	NOT (a OR b), not both	$\overline{a \oplus b}$

At first glance, the tables showing the logic functions of a computer look complicated, and people's eyes usually glaze over, but after you think about it a bit, it makes sense. It should make sense; it is logic, after all! We will carefully go through each of the functions. We need to understand each of them before we can know how a microprocessor or computer works on the inside.

The NOT function

The triangular symbol shown is called a logic symbol. It is not a single electronic component but is a circuit that will perform the function called a NOT *gate*. It is usually inside of a chip. Whatever the input is, the output is NOT. If the input is true, the output is false, and vice versa. In computers, the functions are used for binary and can only have two possibilities. Earlier, we talked about ones and zeros, sometimes called highs and lows. The reason ones are also called highs, and zeros are called lows is because if you were to measure voltages with a meter, a one would have a voltage, and a zero would be near zero volts. We are not too concerned with the exact voltage but are interested in the range in which it falls. Using ones and zeros for an explanation: if the input at x were a one, then the output x with a *bar* above would be a zero. If the input at x were a zero, then the output x *bar* would be a one. Sometimes, it helps understand the operation of logic gates by seeing the inputs and outputs in table form. The tables for digital logic are called *truth tables*. The truth table for a NOT gate is concise.

Input	Output
x	$\overline{x}$
$\overline{x}$	x

The OR function

With the OR function, if a OR b is true (or both), then the output is true. A one on input a OR input b (or both) will cause the output to be a one. The following truth table shows all of the possibilities for ones and zeros in an OR gate circuit.

Input a	Input b	Output
0	0	0
0	1	1
1	0	1
1	1	1

It may help to understand the operation of logic gates by looking at an analogy. An analogy is a comparison to something different but acting in the same way. A good example is how light switches behave in *parallel*. A parallel circuit has more than one path for current to flow. In the wiring diagram below, turning on by closing the contact to either switch S1 OR S2 will allow a current path through the circuit. The logic of this parallel switch configuration is the same as an OR gate.

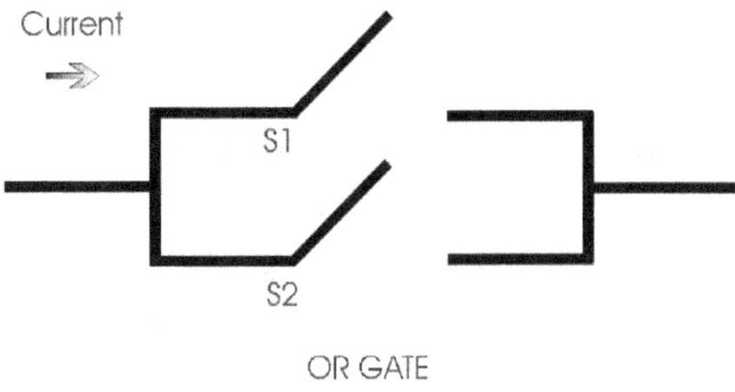

The AND function

For the AND function, both a AND b must be true for the output to be true.

Input a	Input b	Output
0	0	0
0	1	0
1	0	0
1	1	1

A good analogy for an AND gate's operation is for light switches to be in series, one right after the other. For current to flow in the series circuit shown below, both switch S1 AND S2 must be turned on by closing their contacts, to allow a path through the circuit.

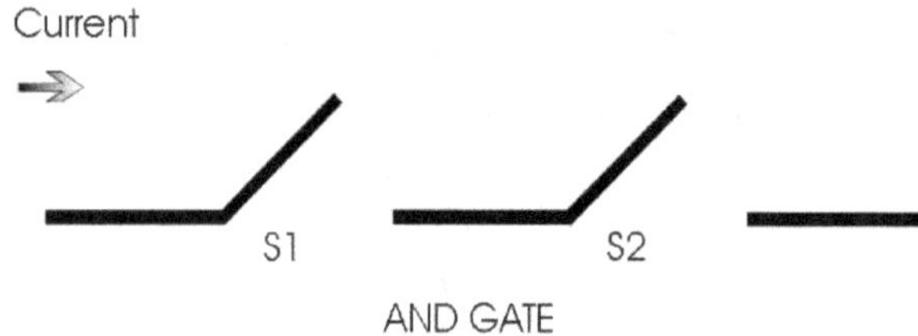

The NOR function

If a OR b (or both) is true, then the output is false. The NOR and the NAND functions get a little tricky, and it helps to use a two-step process when following logic through these gates. For the NOR, think OR-NOT. The NOR function could actually be made of an OR gate followed by a NOT gate. You can use the logic for the OR function, and then just invert the output. The bubble shown in the logic diagram means to invert the logic level after the OR section.

Input a	Input b	Output
0	0	1
0	1	0
1	0	0
1	1	0

The NAND function

If a AND b are true, then the output is false. As with the NOR, it is best to think the logic through in a two-step procedure. Think about how an AND gate works and then invert the output level.

Input a	Input b	Output
0	0	1
0	1	1
1	0	1
1	1	0

Section 4.3. Combinational logic

Earlier in the chapter, we introduced microprocessors and mentioned they are microcomputers. Just as a regular computer has many devices associated with it, a microprocessor also needs additional devices. In previous chapters, when we looked at the parts of a standard computer, we spoke of the computer hardware clock. One of the places it connects is the CPU, which is the processor of a computer. The logic gates that we are describing are an internal part of a microprocessor, but they may also be found in chips associated with it. Logic gates do not need a computer clock in order to operate. When ones and zeros are input, after a short propagation delay in the order of nanoseconds, the result appears at the output. We can connect logic gates to perform complex logic operations. As we described, a NOR gate's logic could be implemented by connecting a NOT gate to follow an OR gate. The logic of a NAND gate is equal to a NOT gate placed after an AND gate. The gate's propagation delays are additive, but this usually isn't a problem since they are in the low nanosecond range. When we design circuits to perform complex logic operations by connecting multiple gates, we call it combinational logic.

<u>Combinational logic example</u>

<u>Problem</u>

We want to design an alarm control circuit for a retail store so that the alarm will sound if the armed switch is on, and any of three door switches are activated. We also want a panic switch so that we can manually sound the alarm.

<u>Solution</u>

There isn't a single logic gate that will perform this complex function, so we will have to design a logic circuit that uses a combination of gates. There are several ways to begin to design the circuit. It might help to put the words into a mathematical form:

> We will call the alarm output x
>
> the three switches a, b, and c,
>
> m is for the manual switch,
>
> and we will call the armed switch d.

The idea is: (a OR b OR c) AND d, OR the panic switch m.
You could write the logic formula as: (a + b + c)(d) + m = x
You could also rewrite the two sets of terms in parenthesis by distributing the d. So, an equivalent formula is: ad + bd + cd + m = x. The first formula seems like it may be simpler to wire. We will draw a logic diagram of both circuits.

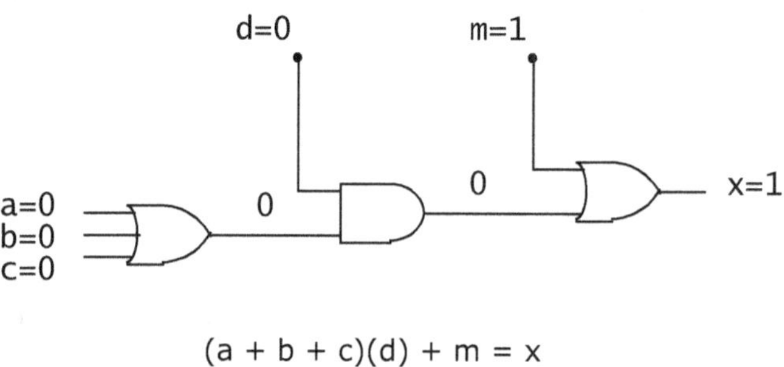

$$(a + b + c)(d) + m = x$$

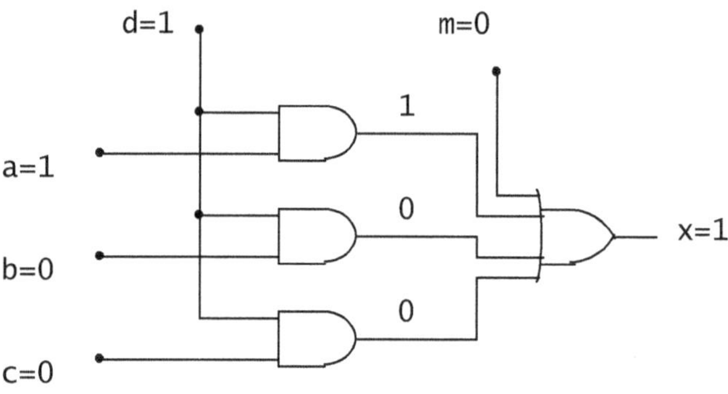

$$ad + bd + cd + m = x$$

We were right about the formula with fewer terms being the simplest circuit to wire together, but other factors are equally important. The more complicated

54

circuit on the bottom would only take two chips to build, whereas the simpler looking circuit on top would take three chips. The two circuits, just like the two math formulas, are equivalent. We cheated a bit in our design and mixed IC technologies, which will be explained in the next chapter. With some slight modifications to our logic diagram, we can provide a work-a-round and will examine this in more detail at the end of this chapter.

Section 4.4. Reverse engineering

It is challenging but possible to identify unknown Integrated Circuits (IC chips) from experimentation. Under test conditions, you could simulate inputs and observe the output. Also, if we had any number of logic gates in combination, we could predict the output logic level with given input conditions. We also could derive the circuit's logic equation and possibly redesign it to be more efficient. In this section, we will do both operations.

To follow the logic levels through a combinational logic circuit with known gates and given input conditions, start on the left and identify the logic level after each gate. Using the gates' outputs on the left as inputs to the gates on the right, you will eventually find the final output level. We will evaluate the sample circuits that we already looked at in the last section.

<u>Unknown logic level output example</u>

The given input conditions are that all doors in the business are closed, a, b, c all equal zero. The burglar alarm switch d is not armed. But someone pressing the panic button m causes the alarm to sound.

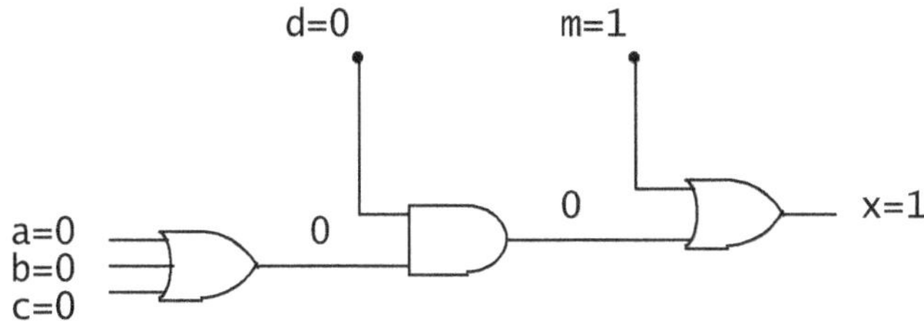

In the circuit from the alarm system shown above, we will do what was just described in the previous section. Starting at the left are the given inputs to the 3-input OR gate. Since none are at logic level one, the output of the first gate is 0. That level is one of the two inputs to the next gate, so the AND gate's output is zero as well, and it is used as an input to the OR gate. Since the m level is one because the panic button is activated, the x output is one. Remember the equation $(a + b + c)(d) + m = x$? When you plug in the input values, you can solve for the output x.

$$(a + b + c)(d) + m = x$$
$$(0 + 0 + 0)(0) + 1 = x$$
$$0 + 1 = 1$$

We were able to verify that the alarm sounded in two different ways. In the first, we followed the logic levels. In the second, we analyzed the circuit through math and found the same answer. The math is called *Boolean algebra*. It is similar to normal algebra, but you are only allowed to have ones and zeros. People that are more *hands-on* may prefer to visualize following logic levels of ones and zeros through the circuit, while people better with math may want to use a formula to find the answer. It's essential never to limit your options, so we will learn both methods. Although, as we will see next, Boolean algebra can get a little wacky at times.

<u>1 + 1 = 1 Boolean algebra proof</u>

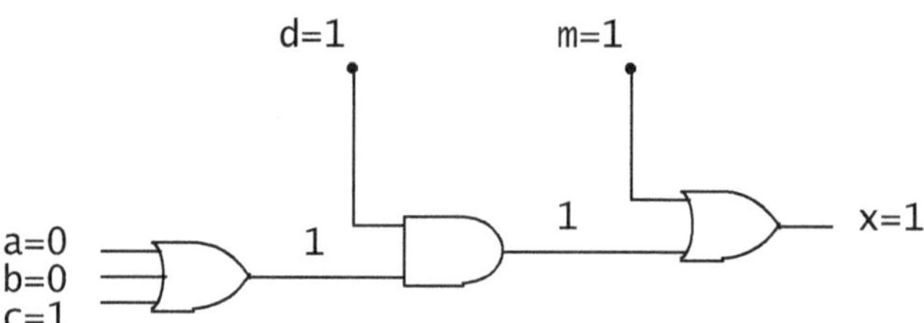

Now, the given input conditions are that the burglar alarm switch d is armed, and someone opens door c. The panic button m is also pressed to sound the alarm. Starting again on the left, since one of the inputs to the 3-input OR gate is 1, the gate's output is 1. The AND gate now has its condition met, so its output is 1. The final OR gate has one on both inputs. The output x is 1. Again using the logic equation (a + b + c)(d) + m = x, we have:

$$(a + b + c)(d) + m = x$$
$$(0 + 0 + 1)(1) + 1 = 1$$
$$1 + 1 = 1$$

If you missed an easy arithmetic problem back in first grade, now you can go back to your teacher and tell them that you were doing Boolean algebra! Next, we'll look at how to develop the logic equations.

<u>Finding a logic equation</u>

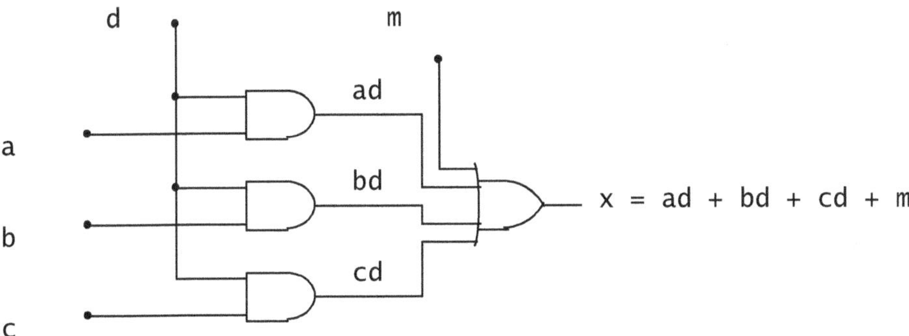

You may remember this circuit as being the second way to design and build our burglar alarm. Both were identical, and even though this one is more complicated to wire, it will use fewer chips. To find the logic equation, we will use the same process that we outlined for finding logic levels. We will start at the left and work our way to the right, writing the math expressions as we go. The operator for OR gates is ┐ , and it is x for AND gates. The top AND gate has the inputs a and d, so the output is ad. The middle AND gate has b and d as inputs, so its output is bd. The bottom AND gate output is cd. Each of the AND gates is an input to the 4-input OR gate, along with the panic button m. Using the OR operator the output x = ad + bd + cd + m.

<u>Building the circuit</u>

Using the old standard TTL logic devices, the AND Gate part number is 74LS08, it is a two-input quad AND gate, which means there are four AND gates in one chip – each gate having two inputs, so the circuit would only need one AND Gate IC. To work within the same IC technology called TTL, we would use a 74LS32 OR gate. It is a quad OR Gate, but we need four inputs? Is it possible to wire two OR Gate outputs into the two inputs of a third OR Gate to make an equivalent circuit, as shown in the diagram? We leave it as an exercise to the reader to draw the logic diagram and decide.

Chapter Four Summary

Digital logic doesn't have issues with semantics like philosophical logic, but both require analytical reasoning. The main logic gates are NOT, AND, and the OR Gate. Adding a NOT Gate after AND and OR Gates gives us NAND and NOR Gates. There is also the exclusive OR (XOR) and its opposite logic gate, the exclusive NOR (XNOR). The small-scale integration gate ICs usually have two inputs and one output and come in a quad package. Devices are available with more than two inputs. When ICs are linked together, it is called combinational logic. Digital ICs can provide a specified output under certain input conditions. The operation of digital logic circuits may be replicated through a programmed microprocessor; however, if the circuit is simple, it may be better to use stand-alone devices.

Chapter 5

Integrated circuit technology

Section 5.1. What's inside the chip

The proper name for a chip is an *Integrated Circuit* (IC). Its name is because many components make up the device and are connected in one assembly. The parts allow for functionality with minimal need for external components. There are still many stand-alone components called discrete components, but ICs can be found in almost every type of modern electronic equipment in today's world. Initially, ICs were only used in low power applications, but now some can handle enough power to drive motors and have even replaced power output components in audio and radio amplifiers. Almost all ICs are made of *silicon*. Silicon is the main component of sand, but the silicon for electronics must be ultra-pure. IC fabrication takes place in clean rooms that are much cleaner than hospital operating rooms. The main component in an IC is the transistor. Transistors were invented in 1947 by Bell Labs and soon replaced *vacuum tubes,* sometimes called electron tubes. Figure 5.1 is a diagram of a triode tube that could part of an amplification circuit. Since many of the modern electronic devices we have today owe their existence to the vacuum tube, let's see how it works:

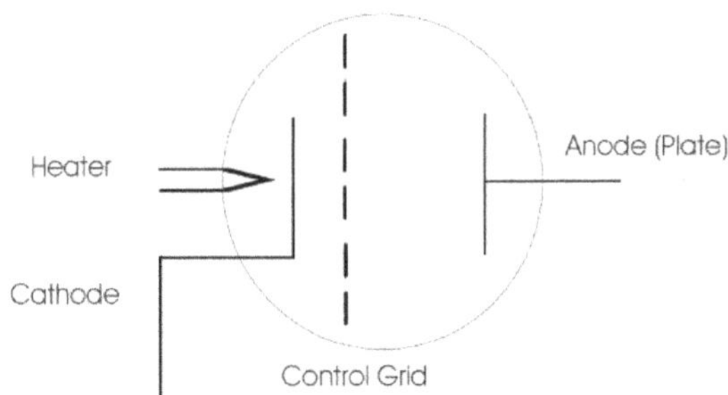

Figure 5.1 Triode Vacuum Tube

Some specialized tubes are still in use today. The older style TV and computer monitors used a picture tube, which is a particular type of vacuum tube. The tube we are examining is a *triode* and has three main elements: cathode, control grid, and anode. The heater provides a high temperature to the cathode and is not considered as a separate element. As shown in the diagram, a positive charge is applied to the anode. The cathode can pull electrons up from the ground potential because they are attracted to the highly positive anode. Remember that opposite charges attract. The anode voltage is usually very high, a few hundred volts in a standard tube to as much as 30,000 volts for the old TV picture tube. With the heater warming the cathode, electrons gain energy and vibrate hard enough to leave the metal surface. The electrons can provide a current flow because the vacuum inside the tube offers no resistance. Because of the negatively charged electrons' attraction to the highly positively charged anode, they move through the tube in that direction. The control grid regulates the intensity of the electron flow. If a more negative charge is applied to the grid, the electrons are less attracted because of the like charge, and they will not travel in as much intensity to the plate. If less of a negative charge is applied, the electrons are allowed to move more freely across the grid to the plate. The plate is connected to external circuitry that eventually provides a path for the electrons to return to ground potential. Depending on the control voltage, the tube just described can act as an amplifier or act as a switch. In computers, the processing is done with devices acting as switches.

The physical size of vacuum tubes is typically about half that of a standard size light bulb, and they are usually made of glass. They may even produce a slight glow and are hot to the touch. In the vacuum tube era, computers were huge and expensive. They also created a lot of heat and required massive amounts of electric power. Remember the ENIAC in Chapter One? After the invention of the transistor, tubes were replaced whenever possible. Transistors don't have a vacuum and are called a solid-state device. Miniature size transistors are the majority of the components that comprise an Integrated Circuit. The early solid-state material used was the element *germanium*. The first IC was developed by Jack Kilby at Texas Instruments in 1958 and was two germanium transistors interconnected together in one package. Germanium has limited applications today and is mainly found in only a few components used for radio communications. Almost all of the first germanium devices have been replaced by silicon. Silicon is more reliable under stressful conditions, and sand is very plentiful.

Before we investigate transistors, a three-terminal device, let's look at a two-terminal device that allows current to flow only in one direction called a *diode*. Diodes are like an electric valve and first became popular as selenium rectifiers, which used two dissimilar metals to change AC to DC. Two element vacuum tubes, also called diodes, accomplished the same purpose and were replaced by the solid-state devices in use today. We know that atoms of good conductors have only one electron in the outer electron shell called the *valance* cloud, and the atoms that are good insulators have a full complement of eight electrons in that region. A semiconductor like germanium and silicon are halfway at four electrons in the valance cloud. For solid-state devices to be functional, however, impurities must be added to the pure semiconductor for the device to have dissimilar sections.

When impurities are added to the pure semiconductor, it is called *doping*. Two different types of materials will be separately added. One will have excess electrons, shown on the left side of the diagram of a diode in Figure 5.2. The new

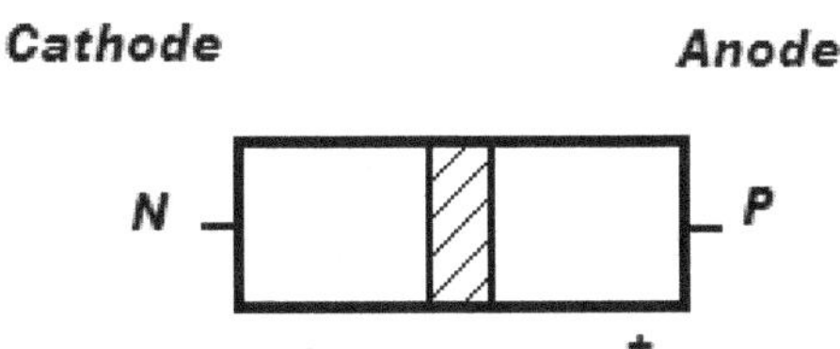

material is called N material. The other side will be doped with a substance that has a deficiency of electrons. The new material formed in the Anode is P material.

When the N material and the P material are put into contact with each other, magic happens! In the area of conjoinment, a *junction* appears that is something of a no-man's-land. It is a barrier junction in which the doped material normalizes back to the pure semiconductor state. It can almost be thought of as a conductive speed bump because a certain amount of force must overcome its resistance for current to flow. This is known as a diode drop. A schematic of a diode is shown in the following figure.

A voltage of 0.3 volts is required in a germanium diode, and 0.7 volts is needed to overcome the barrier junction of a silicon diode. If the voltage is polarized properly across the diode's negative and positive sections, and the voltage exceeds the diode drop, then a current will flow through the diode. If the applied voltage's polarity is reversed, no current will flow except for a negligible amount of reverse current. However, if the backward polarity voltage is higher than the diode's rating, the reverse current will cause an avalanche effect that will damage the diode.

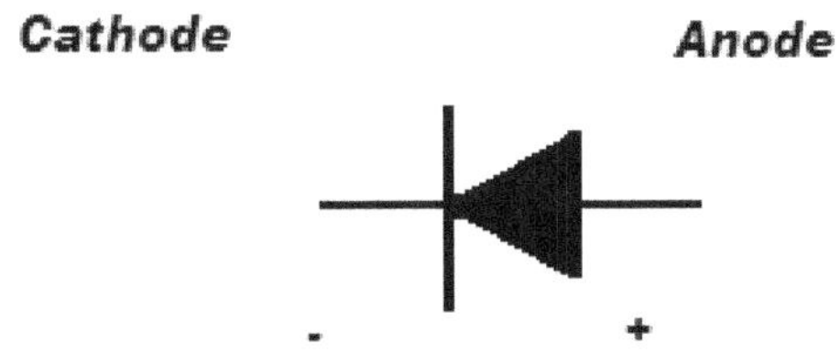

In both a diode vacuum tube and solid-state diode, the elements have the same names. As shown in Figure 5.4, a transistor can almost be thought of as a solid-state version of a triode vacuum tube, but the naming convention is entirely different.

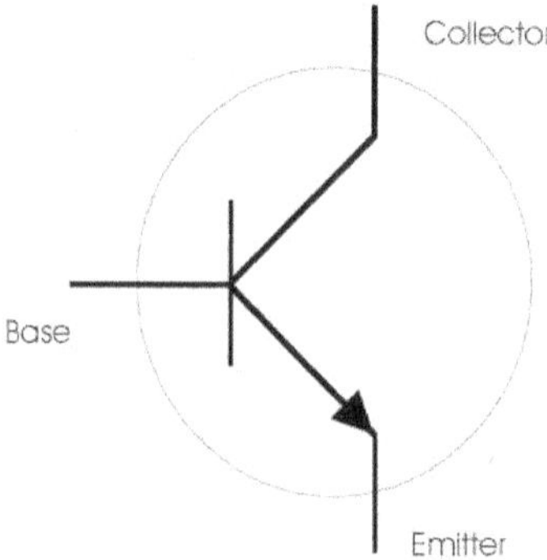

Figure 5.4. NPN Transistor schematic diagram

In the triode tube, the element that was connected to ground potential and supplied the electrons was called the cathode. In a transistor, it's called the emitter. The control element of the tube used a grid. The control section of a transistor is called the base. The anode, sometimes referred to as the plate in a tube, was the positive element that attracted electrons. The corresponding section of a transistor is called the collector.

A transistor is constructed from two back-to-back diodes and has two junctions. In much the same way that the vacuum tube worked, the transistor works similarly but through a different mechanism. The *Bipolar Junction Transistor* (BJT) we are discussing is a current controlled device. It's called bipolar because the junction's N and P sides are similar to the two poles of a magnet. And it's current controlled because a small amount of base current controls a large amount of collector current. Two external circuits are needed to produce two separate loops of current. The same power source could be used for both collector and base, with higher resistive components limiting the amount of base current. There are two types of flow through semiconductors. One is the electron flow, and the other can be thought of as hole flow in which there is the absence of an electron. These two charge carriers move in opposite directions. Bipolar transistors can be constructed as either NPN or PNP. The letters show the type of semiconductor material that is used in each section. NPN means that N doped material is used in the emitter, P in the base, and N in the collector. The NPN configuration is the most common bipolar transistor style. Occasionally, both the NPN and PNP styles are used in complementary symmetry.

62

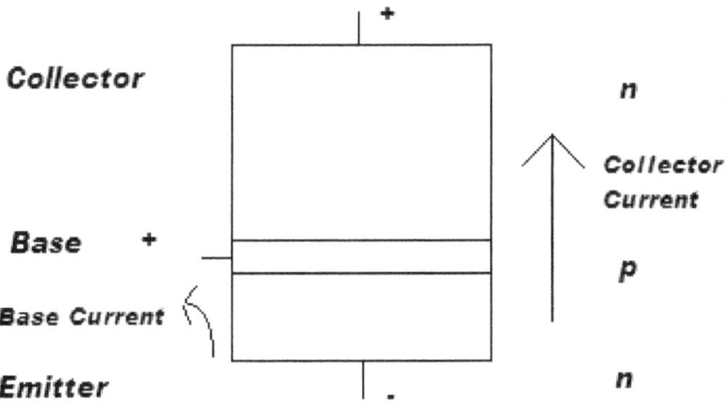

Figure 5.5 Pictorial view of a BJT transistor

In the operation of an NPN transistor, shown in Figure 5.5, electrons are pulled into the base-emitter current loop because a positive voltage is connected to the base lead. Current flows through the base-emitter PN junction like current flows through a diode, but more electrical force is applied to an NPN transistor's collector section. The base region is thin, and because of the higher positive force from the collector, 95 % of the electrons entering the base are sucked out through the collector. The collector sweeps up the electrons, almost like a vacuum cleaner picks up dirt particles. The collector is physically the largest section of the device. Most of the electrons in the collector's N material are concentrated in the area near the positive voltage lead. The area nearest the base region has fewer electrons in the N material. Consequently, it does not impede electrons' migration out of the base and into the collector region. In the area where electrons are concentrated in the collector, the energy causes vibrations and collisions that encourage the electrons to free themselves from the semiconductor material and flow into the metal collector lead toward the positive voltage source.

The bipolar transistors we have been discussing have a few limitations and have limited application for use in the very densely packed ICs of a computer. The main prohibitions are a somewhat larger size, more expense, and greater power consumption than the *transistor's unipolar style* called a Field Effect Transistor (FET). Bipolar transistors are very fast at switching and used where speed is a predominant consideration. They were used in the *cache* memory of computers but have mostly been replaced in modern PCs by field effect transistors. The cache is a small, fast type of *RAM* in a computer where data is likely to be utilized by the CPU rapidly. L1 level one cache is located inside the CPU, with L2 usually located within the same IC. L3 cache may be found near the CPU IC on the motherboard. RAM stands for Random Access Memory and is the working memory of the computer. It is used for short-term storage of data and is said to be volatile because the data is lost when the power is shut off. The RAM memory that most people think about is the memory that can be upgraded and is located on memory modules called DIMMs. The ICs on the DIMMs use Unipolar memory transistors that utilize voltage field-effect rather than current control for transistor operation. Unipolar FET transistors many times use Complementary Metal Oxide Semiconductor (CMOS) configurations. CMOS uses two FETs, one made of N, and one of P material for their main channel sections. We illustrate the P material FET's operation in Figure 5.6.

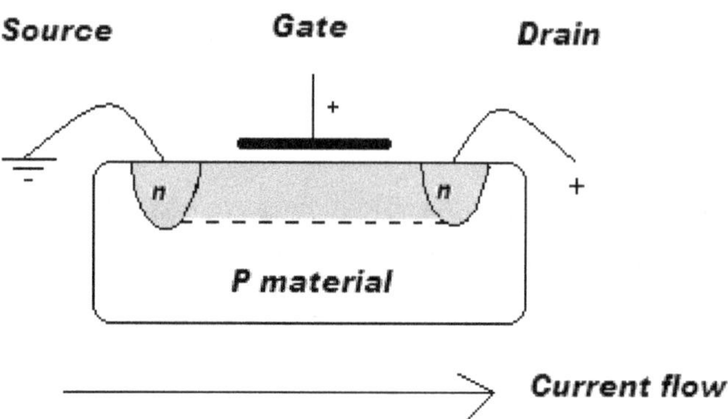

Figure 5.6 A field effect transistor

We again have different names with which to contend. The source (shown to the left) is like the emitter of a bipolar transistor, the gate (center) is similar to the base, with the drain similar to a collector. In many FETs, the source and drain are interchangeable. Unlike the base current used bipolar transistors to control the main collector current flow, the FET gate voltage is insulated from the semiconductor material. There is no control current flow, but only a voltage force. At the source and drain positions, a small area called a *well* is doped with the opposite material. In our example, N material is diffused into the wells. When a positive voltage is applied to the gate near the P material between the N wells, some of the free electrons in the P material migrate toward the gate. The electrons between the wells move because of the attraction of unlike charges, and in essence, form a long channel of N material between the wells. Current can now flow through the single channel because the barriers of the different materials no longer block it.

The original field-effect devices were quite a bit slower than bipolar devices, but rapid advancements have been made, and the disparity in speed is not as much as an issue today. Field-effect transistors and especially densely packed CMOS ICs have the drawback of being extremely sensitive to static and can be damaged if not handled correctly. As the packing densities increase in ICs, the geometry has become very tiny, with the insulating areas as small as a few atoms in thickness. You should always observe Electro Static Discharge (ESD) precautions before handling field-effect devices. When working on a computer or other equipment devices, an easy way to eliminate any static charge you might have picked up is to touch the bare metal chassis before handling ESD sensitive devices. Anti-static mats and wristbands should be used whenever possible.

Section 5.2. How ICs work

ICs are extremely easy to work with, and they are like building blocks that can be assembled to accomplish amazing things. The connection pins on an IC are inputs, outputs, and pins to power the device. It's unnecessary to know what the circuit looks like on the inside of the IC, but we should have a general understanding of how the transistors work. We know that transistors act as switches in digital devices, so let's see how an output can either be at a high or low logic level. We know about resistors, diodes, bipolar and unipolar (FET) transistors, and that is what is inside digital ICs. A bipolar IC technology that has been around for a long time is called Transistor-Transistor-Logic (TTL). The output of a TTL device is described in Figure 5.7.

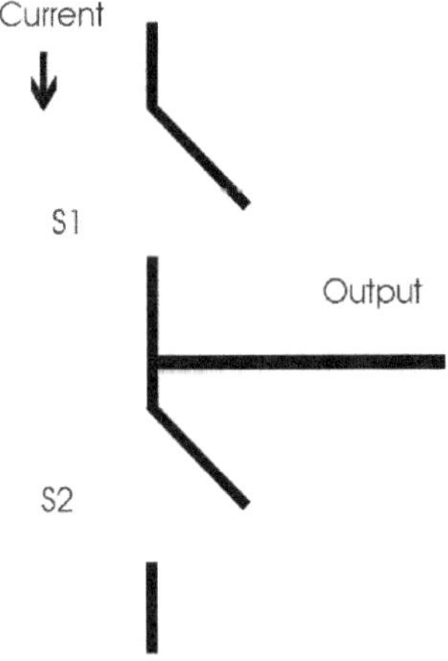

Figure 5.7. TTL Totem-Pole design

In a TTL output, there are two transistors in a totem-pole design that act as switches. The top transistor is connected to a positive voltage, and the bottom transistor is connected to ground or lower potential. Just as in the drawing of Figure 5.7, the output is taken between the two of them. Other transistors control the totem-pole bases and allow only one transistor at a time to turn on. In actual operation, one of the switches in our drawing would be closed and the other open. When a switch is closed, it is on and makes a connection to allow current to flow. If the output of the device were producing a high, switch S1 in our drawing would close, and S2 would be open. The positive voltage would be felt at the output. This is called *sourcing* current, since any device connected to the output could provide a path to allow current to flow out of the IC. In the other state, the output is low, switch S1 in our drawing would be open, and switch S2 closed. Ground potential would appear at the output pin. This is called sinking current since any device connected to the output could provide a path to allow current to flow into the IC.

In TTL devices, the amount of integration is quite low by today's standards. The venerable part number 7408 is a very common quad AND gate IC. There are four AND gates in the package, and you can leave any unused gates disconnected. Six transistors form the totem-pole and control circuitry for each AND gate. In the 7408 IC, a total of 24 transistors are contained in the package of the four gates. This is an example of small-scale integration. It's more likely that you may deal with the 74LS08, which is a more modern version of TTL. The LS stands for *Low-power Schottky*. The device consumes less power and runs faster than the original version. Other versions feature improvements over the original part, but they all remain small-scale integration devices. Small-scale integration has not been eliminated as technology has progressed because only a few AND gates may be all that is needed for a project. Logic gates may be part of a system that connects to a microprocessor or a computer. In the latest computers, the CPU contains upwards of a billion transistors in one single package. As Giga-scale integration expands, computers can operate quicker and run more intensive applications. One of the founders of Intel made an interesting observation, known as Moore's Law. He said that the number of transistors that can be fabricated in an IC would double approximately every two years. So far, Moore's Law has held true. Many futurists believe that we are coming close to the limit for miniaturization with our current methods, and a new process will need to be developed.

Figure 5.8. EPROM Memory IC

The current process used to construct the miniature electronic circuits inside an IC uses a system similar to photocopying and printing. It is called photolithography and involves a repeated process of developing an area of silicon to accept either N or P dopants and then applying metal interconnects and insulation. Finally, the silicon IC, called a die, is packaged, labeled, and tested. In the above picture in Figure 5.8, a die is visible, as are the wires connecting to bond pads that attach the external pins.

The concept of reducing the size through photography is not as amazing as is the complexity of manufacturing these devices at the nano-scale level. A nanometer is a billionth of a meter, and current die fabrication is at single nanometer geometries. As distances shrink to these tiny sizes, Newtonian physics principles are overwhelmed by quantum mechanics, and devices do not always behave the way we would normally think they would act. Some experts believe that within 5 to 10 years, we will reach the limit of the photolithography method and may need to begin processing data through the use of photons, the spin of particles, or some other method yet to be discovered. We will explore the realm of possibilities for data processing and information exchange later in this text.

Chapter Five Summary

Digital electronic computers first contained vacuum tubes and relays acting as switches. After the development of solid-state devices, the transistor replaced the large and inefficient tube, and as the transistors were miniaturized, they became part of integrated circuits. Complementary Metal Oxide Semiconductor CMOS technology is mainly in use today, but TTL is the old standard technology used for digital devices. CMOS is static sensitive, and ESD precautions should always be followed to avoid damage to the parts. TTL devices have a totem pole output construction where either one of two transistors conducts. On a high output, current flows out of the IC, and it is said to be sourcing current. Conversely, when an output is low, the device is sinking current. Outputs must never be connected to other outputs, or excessive current flow could result since one could be sourcing while the other may be sinking current.

Chapter 6

Programming a computer

Section 6.1. Manipulating the ones and zeros

Human reasoning and language are very complex. When we program a computer, a special interpreter program translates from the human world to the computer realm. Ultimately, our human reasoning and language must be converted into ones and zeros since that is how our computers process information. In the last few chapters, we learned that it is easy for the electronics in our logic circuits and computer processors to deal with voltage levels that are either high or low. The binary system only contains ones and zeros. It is known in mathematics as the base-2 system. Computer programmers and users don't deal much with this low level of information, but people working on the hardware side of things should understand it. The math rules are very straightforward, and the reasoning follows along with how we have been dealing with numbers since the days of elementary school.

<u>Explanation of the decimal system</u>

If Martians were to land a spaceship in your backyard and ask you how the number system worked on Earth, you might explain it this way:
Our number system uses ten characters, zero through 9. Each character is associated with a weighted column. We have a decimal point, and every number to the left of the point has a value that gets larger as each weighted column progressively moves farther away from the point. Numbers located to the decimal point's right are less than one unit and get progressively smaller as the columns move farther from the point.

With that explanation, the Martians would reenter their rocket ship and quickly leave the Earth! But that is exactly how our number system works. It is called the decimal or base-10 system. We never give it a second thought, but at first glance, it may seem very confusing. It has been a long time since learning this in first grade, but it should give you an appreciation of the amount of patience that elementary school teachers must have. In explaining the weights of columns in the base-10 system, it may help to show it in chart form:

10^2	10^1	10^0	10^{-1}	10^{-2}
hundred	ten	one	tenth	hundredth

• Decimal point

In working a problem in this system, remember the rule that ten characters may be used and are the numbers 0 through 9. A number larger than 9 would need a carry up to the next higher weighted column.

Problem

Using the base-10 number system's weighted column concept, find the total number for the characters in the following order: one, nine, two. They are located directly to the left of the decimal point.

Solution

Put the characters in the appropriately weighted columns. Multiply the value of each character by the column's weight, and then sum the column values to find the total value.

10^2	10^1	10^0	10^{-1}	10^{-2}
hundred	ten	one	tenth	hundredth
1	9	2	•	

The answer is: $100 + 90 + 2 = 192$

The answer seems ridiculously simple: 1, 9, 2 located in that order to the left of the decimal point equals 192. But we will use the same concept to understand what the base-2 numbers mean in a computer. The base-2 system works the same way as the decimal system except the characters are 0 and 1, and the base 2 number replaces base 10 in the column weights.

2^3	2^2	2^1	2^0	2^{-1}	2^{-2}	2^{-3}
eight	four	two	one	$\dfrac{1}{2}$	$\dfrac{1}{4}$	$\dfrac{1}{8}$

Problem

Using the weighted column concept for the base-2 number system, convert to base-10. The binary characters are in the following order: one, zero, one. They are located directly to the left of the decimal point.

Solution

A way to restate the problem is to say, convert 101_2 to the decimal system.

Notice the small number 2 after 101 in the previous sentence. It is called a *subscript* and tells you that the number 101 is a base-2 number. For decimal numbers, it is understood that the subscript is 10, and it is not usually shown. To solve our problem, we use a similar operation to when we found the sum of the column weights for the base-10 system. One of my past students had an excellent way of explaining the process. They said to think of the ones as turning columns *on* and zeros turning columns *off*.

2^3	2^2	2^1	2^0	2^{-1}	2^{-2}	2^{-3}
eight	four	two	one	$\dfrac{1}{2}$	$\dfrac{1}{4}$	$\dfrac{1}{8}$

1	0	1

The answer is: 4 + 0 + 1 = 5.

101_2 = 5 in the decimal system.

Binary numbers can contain strings of ones and zeros that can become quite large. Two higher base numbers systems use more characters and convert nicely to-and-from binary. The first is octal, which is base-8, which uses the characters 0 through 7. The octal system is not used much anymore. The other that remains popular is called the hexadecimal or base-16 system and uses the symbols 0 (zero) through F. The hexadecimal system is handy for low-level programming. It allows us to represent the ones and zeros in-groups of four. It enables us to deal with data in a concise form that can easily be converted back-and-forth to binary. As we mentioned, the base-16 system uses 16 characters per weighted column. Our decimal system stops at 9, so letters are used to represent values above 9 in the following way:

$$A = 10$$
$$B = 11$$
$$C = 12$$
$$D = 13$$
$$E = 14$$
$$F = 15$$

Converting back-and-forth to binary is simple since each hex number contains four binary digits called *bits*.

Examples of hex-binary conversions

The operation is to represent a hex character as four bits, and vice versa.

$FF_{16} = 1111,1111_2$. $101_2 = 5$ $A1_{16} = 1010,0001_2$ $1110,0001_2 = E1$

Notice that each hex character is directly associated with a group of four bits. It is more challenging to convert back-and-forth between hex and the decimal system, and you may wish to use a chart as we did in the previous problems.

Problem

Convert $1AF_{16}$ to the base-10, decimal system.

Solution

Using the base-16 weighted columns, we can find the answer in a similar fashion to the other conversion problems that we looked at earlier:

16^2	16^1	16^0	16^{-1}	16^{-2}
256	16	1	$\dfrac{1}{16}$	$\dfrac{1}{256}$

1   A   F   •

The answer is: 256 + (16)(10) + (1)(15)

256 + 160 + 15 = 431

Some calculators allow you to easily convert between bases, especially between the bases we discussed useful for computer engineering. You might need to change modes on some calculators, or you may have direct keys to make the conversions. Hex can be used in programming many microprocessors and IoT devices, but for typical operations involving PCs, you will generally program in a higher language level that mimics English.

Section 6.2. HTML

Standardization has been an instrumental factor in the rapid proliferation of computers across the world. Standardization is the norm for both hardware and software and has helped reduce the cost of equipment. There are a few specialized machines and software packages, but generally, most systems conform to the same standards. Over the years, many different programming languages have come and gone. In my opinion, the three most useful languages are Python, C++, and Javascript. The languages are considered to be high-level programming languages. As we mentioned in the last section, while it is possible to have a great deal of control over a computer and use low-level language for programming, it is very cryptic and tedious. High-level programming languages use almost everyday terminology, although it is highly structured and formatted, so there is no question about the meaning of the words. The formatting is known as *syntax*. Before we see how to program a computer, let's examine a type of programming language mainly used for designing Web pages.

The language that displays Web pages on your computer is called *Hyper Text Mark-up Language* (HTML). It works with your Web Browser. Although many web pages contain javascript and other add-ons, basic HTML is a straightforward way to explore computer programming. HTML uses a system of commands called tags that instruct the browser how to display information. Tags are written inside of brackets such as these < > and not displayed on the Webpage.

Most Web browsers give you the option to see the HTML code. You should be able to see the code of a Web page by going to the browser menu and selecting *view source*. You might also be able to right-click on a background area of a Web page and choose to view the source. Many Webpages use very involved source code due to specific positing of the content and for interactivity, but simple source code may look something like this:

```
<HTML>
<HEAD>
<TITLE>My Web Page</TITLE>
<META NAME="keywords" CONTENT="webpage, programming, practice">
<META NAME="description" CONTENT="This is my page.">
</HEAD>
<BODY>
<H3>Welcome to my Webpage</H3>
< BR>
<IMG SRC="mypicture.jpg" ALT="my picture">
<P>Thanks for visiting. Stop back again</P>
</BODY>
</HTML>
```

The code just shown would display a Web page like Figure 6.1.

Figure 6.1. Webpage example

Closely examining the source code:

```
<HTML>
```
Tells the browser it's an HTML document.

```
<HEAD>
<TITLE>My Web Page</TITLE>
```
Displays the title information at the top of the browser.

```
<META NAME="keywords" CONTENT="webpage, programming, practice">
<META NAME="description" CONTENT="This is my page.">
```
Meta information is used only by search engines.

```
</HEAD>
```
Stops the heading section

```
<BODY>
```

Starts the main part of the Web page.

<H3>Welcome to my Webpage</H3>

The H3 displays the words as a headline. H1 is the largest, and H4 is the smallest font.

< BR>

A break to add some empty space.

<IMG SRC="mypicture.jpg" ALT="my picture">

This tag displays an image. (You must have a picture of the same name in the same directory as the HTML code, otherwise you could use a different instruction that points to the location of the picture. The ALT instruction displays the text *my picture* when the image is moused over.

<P>Thanks for visiting. Stop back again</P>

This is a paragraph of small font.

</BODY>

</HTML>

Stops the body section and the HTML.

You can make your own Web pages that will display in a browser on your local machine. You can also upload the HTML to a Web server so that it shows on the Internet. Some Web page design programs will write all the code for you, or you can do it yourself in a text editor program like the Microsoft text editor called *Notepad,* which has come bundled with the Windows operating system for many years. It is located in the *accessories* category of Windows. After you type the code in Notepad, you must save it as a .html file for display as a Web page.

It is debatable if HTML is computer programming. There are many programming languages embedded in Web pages and run as applications. In the 1990s, Internet users were quite happy to *surf the net* by going from Website to Website for no particular reason. Most users were on slow dial-up connections, and Websites had to be pretty simple to load in a timely fashion. There was little use of video or sound, and even graphics and pictures had small file sizes. Download speed is no longer an issue because most people have high bandwidth connections. There is much higher quality to Website design today, and many people depend on them to be both functional and entertaining. The Internet evolved into what has been called the 2.0 stage, with a great deal of interactivity now taking place. The languages that we will look at next can be used for this purpose. The languages can produce programs that can process data to solve complex problems by performing calculations and making decisions.

Section 6.3. Programming code

Computer programming is problem-solving. The very first step is to understand the problem thoroughly. If you are working with other people on a project, time is well spent at the initial stages to have complete agreement on the problem. Be very careful of *scope creep*, which is when additional features are added to a project along the way. Scope creep may not allow you to finish a project on time and within budget. In program design, you must make sure to keep the end-user in mind and make the program as easy to operate as possible. Before even thinking about the program code, it may be best to design the user interface.

We will explore the Microsoft Visual Basic programming environment for the remainder of this chapter. It is a program that allows you to make programs. A free version called Visual Basic Applications is part of the Microsoft Office suite. There is an excellent book available called *Fun with Office* that explores VBA programming. It is one of my favorite books that I wrote. For our example project, we want to create a simple program that will give the user a message when they click on a button. If you have Microsoft Office programs like Word or Excel, you can create this project. You must first add the developer tab to the menu, and after clicking it, go to Visual Basic. VB is an *object-oriented event-driven* type of programming language, as opposed to *procedural* languages, which do not allow for as much spontaneous interaction from the user. For our message program, we will first design the user interface shown in Figure 6.2.

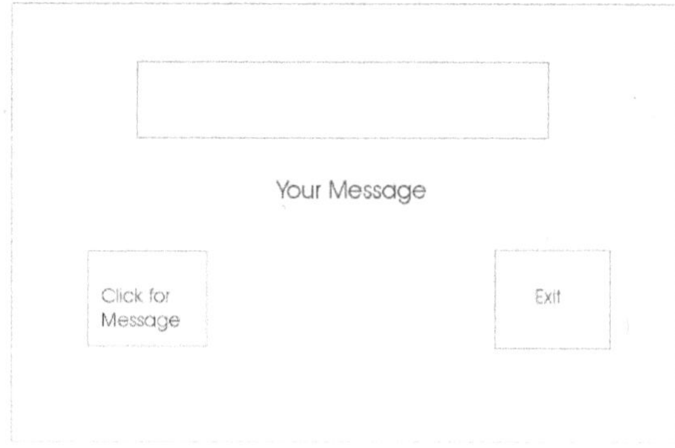

Figure 6.2 Sample user interface

There are two buttons that the user can click. One will display the message, and the other one will exit the program. The message will display in the box above the text labeled "Your Message." The message will simply say, "hello." Visual Basic makes it very easy to construct the user interface, which is called a *form*. When you begin programming, you start with a blank form in the work area, and you add the buttons, called *controls*, by selecting them from the toolbox located on the left side of the screen. You do the same to add the textbox to the form where the message will appear. Text that will always appear on the form is called a *label*, and it also comes from the toolbox. In our project, we want the words "Your message" to always be displayed on the form. Care must be taken in naming every object added to the form since you must refer to it by the exact name in the code. You list the name of each item in the properties area. An example of the Visual Basic programming environment is shown as a screenshot. This program was designed to calculate sales tax. The screenshot in Figure 6.3 shows the form at design time inside the Visual Basic IDE (Integrated Development Environment).

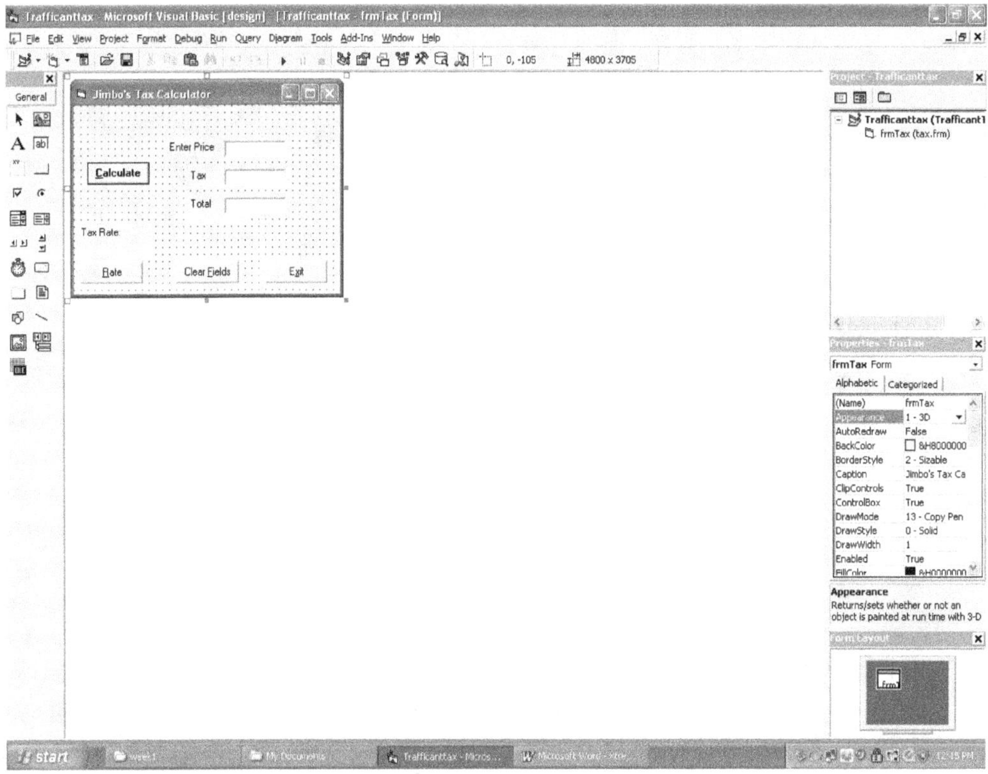

Figure 6.3. Microsoft Visual Basic programming environment

The programming environment looks challenging to use, but it is very intuitive, and you can start writing simple programs in a short time. There are many complex features available in Visual Basic, and with some practice, you can make very robust programs. I think VB and VBA are great ways to begin learning about coding.

Even the most experienced programmers will sketch out their solutions as a roadmap or jot down their ideas in words before writing the computer code. A simple type of sketch is called a flow chart, as shown in Figure 6.4. Our message program example is quite simple, and a flow chart is not really necessary, but flowcharts can be helpful as programs become complex.

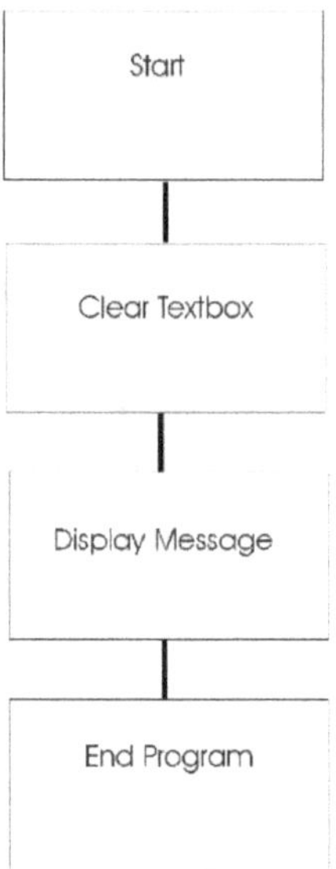

Figure 6.4 Flow chart for the Message Program

Another helpful way to begin to program is to write a pseudocode program. It is an intermediate language between the way that you speak and the formal computer programming code. There is no special format or syntax. For our message program, it might read something like this:

Begin program.
Clear textbox.
On message button click, display message.
End.

The code for Visual Basic is modular and appears in the design environment in sections called subroutines. There is no particular location order for the subroutines. In our project, we named the textbox *txtMessage*, the message command button *cmdCommand*, and the exit button *cmdExit.* To begin coding, you just double-click the control and type into the code screen that appears. In our program, we first double-clicked an empty spot on the form and typed *txtMessage* = "". The open and closed quotation marks are called a null string. It will clear the textbox every time the program starts. The next code we entered was for the command button to display the text. It is shown in the following code listing. Text typed inside of the quotation marks will be displayed when a user clicks the button. Finally, the Exit button was coded to end the program. Visual basic automatically enters the first and last lines of each of the subroutines. Below is how the complete code looks:

```
Private Sub Form_Load()
TxtMessage.text = ""
End Sub


 Private Sub cmdMessage_Click()
TxtMessage = "Hello World"
End Sub


Private Sub cmdExit_Click()
End
End Sub
```

That's all the code you need to have for the computer to say "hello" to the world. This is a very simple program, but the same procedure would be used for more complicated programs. Other programming languages will have differences in syntax, but much of the basic structure and logic will be similar. Old programming languages that were procedural, such as Fortran and Cobol, are now not used much. Procedural languages run their lines of code sequentially, and the users have minimal interaction. They are difficult languages for programmers to debug. It is very seldom that code does not need to be debugged and revised. Usually, there are errors that must be fixed before a program will work correctly. Most current object-oriented languages make it easy to find syntax errors. In a program like Visual Basic syntax errors will light-up in red. It is more difficult to catch logic errors, but with modular design, each section can be tested and debugged separately. Many object-oriented event-driven languages like Visual Basic allow you to debug the operation by stepping through the code and stopping at breakpoints added for debugging purposes.

Chapter Six summary

There are three main numbering systems we work within when dealing with computers. Normal life utilizes the base-10 decimal system, but computers are binary and use the base-2 system of ones and zeros. The base-8 octal system uses three bits, and the base-16 hexadecimal system uses groups of four bits to represent binary numbers in a more manageable way. HTML is the language for webpages on the Internet. Other computer coding languages are sometimes used within HTML to allow for more user interactivity. HTML tags control the layout of displayed Webpages, and the code is contained within brackets <> and not seen by the viewer. It can be seen by viewing the source code. Visual Basic (VB) and Visual Basic Applications (VBA) are object-oriented event-driven, and provide a high degree of user interactivity as opposed to older style procedural languages. In any design process, it is always helpful to totally understand the goals of a project. Before actual coding takes place, it's best to sketch a flowchart or develop pseudocode.

Chapter 7

Specialized computers and programs

Section 7.1. Computer speed

If you use a computer for gaming, you want a speedy computer with exceptional video processing and extra memory. Some people will buy a high-end gaming computer off-the-shelf, while some may want to pick all of the parts and assemble it themselves. Either way, machines designed for gaming are more expensive than ordinary computers. Most motherboards have onboard video processing, but a gaming PC may have a separate video card containing a graphics processer and a massive amount of RAM on a high-speed bus for better video. The video card will have and a specialized video processor with fast throughput. A high-quality video card can be costly. For enhanced performance, all of the parts of a computer need to be optimized, but this is especially true of the video processing circuits. Earlier, we mentioned that bottlenecks in a computer were the hard drive and buses. This can be observed in the long load times for intensive games on PCs. The data can run at a fast rate once it is loaded into the RAM working memory, but performance is noticeably slow as the data is read from the hard drive. A great deal of RAM is needed on the motherboard so that the computer will seldom need to access the hard drive once in operation. Many people prefer game consoles over PCs. Game consoles are an example of specialized equipment. They have been designed specifically for enhanced audio and video processing. The term for the structure of a computer is *architecture*. Just as in civil engineering, where architecture refers to how a building is made, computer engineers will design a computer or gaming machine's hardware architecture.

The latest video game software is very large and hardware intensive. If a modern software package were used on an older PC, the performance would not be very good. Teams of people usually work on different program code sections to develop today's robust computer software packages. This type of project management is called *concurrent workflow*, or *parallel* flow. When a project is very large or time is of the essence, people in different fields of expertise will simultaneously work in cross-functional teams to expedite the job. Just as in the development of software where the parallel workflow speeds up the development process, the data can be processed in parallel for the most efficient operation inside the computer. We will precisely analyze the electronic circuitry that makes this kind of high-speed processing possible in the next chapter.

The computer operating system that most people are familiar with is Microsoft's Windows. They designed their latest operating system and newer applications to take advantage of parallel processing, having gone to processing 64 bits simultaneously. For high-speed throughput, both software and hardware must operate synchronously to allow for high speed concurrent parallel processing. That makes the multi-core processors found in the latest computers outperform single-core processors that run at higher clock speeds.

Design engineers at Intel were seriously considering adding liquid cooling to their processors at one point in time because of the substantial heat generation at extremely high clock frequencies. But, they decided to implement the *multicore* processor structure. The Sony Corporation had a lot riding on the development of the Playstation 3 game console. Its introduction to the marketplace fell way behind schedule because of the innovative engineering of the Blu-Ray DVD and new processor architecture development. It introduced a processor with eight cores. It's almost like having eight independent machines all working in tandem.

Section 7.2. Embedded processors

In much the same way as a gaming machine has a unique video board, a microcontroller may be built-in to a piece of electronics equipment used to function industrially. This type of processor is said to be *embedded*. An Arduino is an excellent example of an embedded processor. The uses for an Arduino, as shown in Figure 7.1, are just about only limited by the designer's creativity.

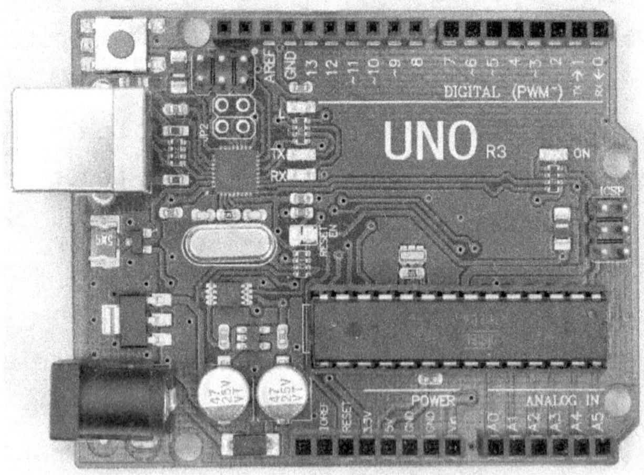

Figure 7.1. Arduino Uno

There is an active group of Makers numbering over 30 million worldwide, designing projects with the Arduino as the microcontroller. The Arduino Uno is a popular development board containing the Atmel ATmega328P microprocessor. It runs nearly a thousand times slower than a PC or Raspberry Pi, with a clock speed of only 16 MHz. It has a working memory of just 2K of RAM, and 32K of Flash long-term memory to store a program, but the low-cost and versatility make it the perfect choice for many Maker Movement projects. Advanced versions are also available that run at much higher speeds, and some even feature dual-core processors. Some projects utilizing Arduinos are quite important and complex. During the Coronavirus Pandemic, a few low-cost open-source ventilator designs used Arduino boards as the controller. The difference between a microcontroller and a standard computer is that the controller is specifically designed to analyze inputs and generate outputs. It has one program that continually loops. There are add-on accessories called *shields* that can increase the functionality of an Arduino. There are shields for GPS location, connecting to the Internet, and even interfacing to operate high power devices.

The Arduino uses a friendly version of C++ for programming. In our example project, we will sample a sensor so that an LED will illuminate for three seconds when activated. We are using a switch with an active-low connected to pin 8 as the input, and the self-contained LED connected to pin 13 on the development board as the output. The Integrated Development Environment IDE is available as a free download from www.arduino.cc.

```
// 3 second delay output program on sensor activation
const int LED = 13;
const int button = 8;
boolean buttonPush;
void setup( ) {
  pinMode (LED, OUTPUT);
  pinMode (button, INPUT_PULLUP);
}
//now comes the main loop
void loop( ) {
  buttonPush = digitalRead (button);
  if (buttonPush == LOW) {
    digitalWrite (LED, HIGH);
    delay (3000); // 3 seconds
    digitalWrite (LED, LOW);
```

        }

    }

The two backslashes are for a single line of documentation, which is ignored by the program. It is always a good habit to provide some documentation to make long programs easier to debug or modify later. Visual Basic and most programming languages declare the data type of the variables. Putting the declaration section at the top of the code is considered a general public declaration and will be upheld throughout the program. Local variables may also be set in different areas and can be changed for each section. The code syntax is precise, and there is no spellcheck! We declare the variable LED in our code to be a constant integer (meaning its value is unchanging), and we associate it with pin 13. We do the same for the button and associate it with pin 8. The pin connections are called headers, and there is a small gap between pins 7 and 8, so they are easy to find quickly, but the pin choices are somewhat arbitrary. In typing the code, the case of the letters is considered and must be consistent.

Next is the setup section, where inputs and outputs are defined. The *pullup* term in the code enables an internal resistor tied to a high logic level. Without any action, the level will be high at the input, but when a button makes contact to ground, the input becomes low. The low logic-level will active our LED later in the previous code listing.

After the setup section runs, we have the main loop. The first part of the code only runs once, but the main loop will keep repeating. The first thing we do in the loop is to read the button's logic-level. If it's low, then the section of code within the curly braces will execute. If the button is pushed, a High logic-level is written to the LED, and it illuminates. The delay is the number of milliseconds to pause before a Low logic-level extinguishes the LED. As the loop circulates, the button condition is continually tested.

This example program can be built-out in many ways where the controller acts to sound alarms, power equipment, or provide any number of outputs as an input sensor's logic-level is changed. The Arduino IC also contains an Analog to Digital Converter (ADC). The output can respond to analog voltage level inputs on that sequence of pins or high and low digital logic-levels.

Section 7.3. PLCs

It may be strange to learn that not every modern computer needs to run at top speed or use a processor with multicore architecture. In fact, there are times when you may need to have a substantial delay. This is especially true in data acquisition, such as when switches and relays close contacts to make a connection. When electrical contact is made, it does not immediately go from the *off* to the *on* state. There is a short period with sort of a flicker, called a *bounce*. It is far too fast for the eye to see, but the logic circuits in a computer may respond erroneously by interpreting the fast bouncing voltage as multiple signal changes. The logic circuits that can operate in the nanosecond range would need to have their response time substantially slowed down to the millisecond range to interface with mechanical switches. Even with the slow down, keep in mind that a millisecond is one-thousandth of a second. That's plenty of time for many industrial applications.

An example of an industrial application might be counting cans of soda pop passing a point on a conveyer belt and activating machinery when 24 cans are ready to pack in a case. A specialized computer specifically designed for industrial applications like this called a Programmable Logic Controller (PLC) is shown in Figure 7.2.

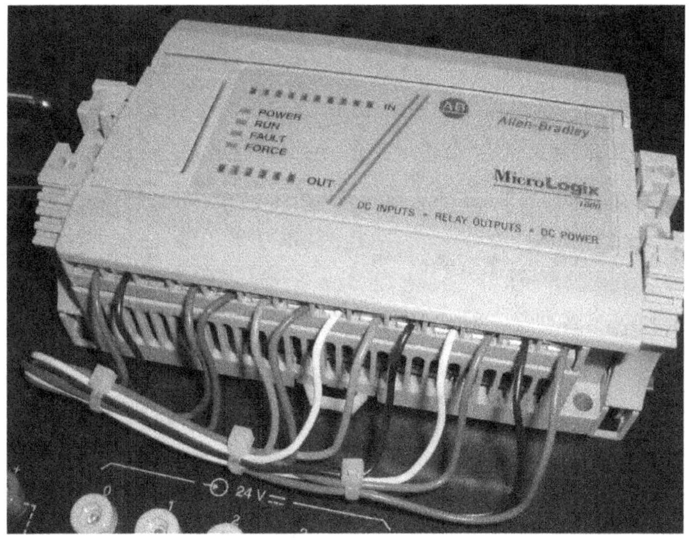

Figure 7.2. Programmable Logic Controller (PLC)

While a PC could be used to control industrial automated equipment, normal computers are a little too generic and lack the specialized functionality of a PLC for the required tasks. Normal PCs can be used in a multitasking environment allowing you to look at Websites, email, and play music, but a PLC can only run one program at a time. In a PLC, there are many input and output *ports* where external sensors and actuators can connect, for the program to interact with the outside world. A port is another word for connection, and regular computers aren't designed to have sensors and actuators connected to them. PLCs may have as many 128 separate I/O ports, and may even be expanded if more are required. As a specialized piece of equipment, the PLC is optimized to perform control tasks in an automated manufacturing environment. They are very reliable, with limited moving parts. They use flash memory in their CPU modules, which is the type of memory similar to the flash drives that have become very popular. The three main parts of a PLC are CPU, I/O modules, and power supply. Since these types of specialized computers are usually running unattended, there is no computer monitor or keyboard. The program is usually developed on a laptop and then downloaded to the PLC. They also can be networked together and monitored in real-time. The programming is as specialized as the PLC; it is done in a visual language called *ladder logic,* which we look at more closely in the next section.

Section 7.4. Ladder logic

In the early days of the personal computer, the operating system that was popularized by Microsoft was called the Disk Operating System (DOS). DOS was in use before Windows was developed and used a text-based environment. In that era, you almost needed to know how to program a computer just to be a computer user. The popularity of the Windows OS was mainly due to a Graphical User Interface (GUI). Sometimes computer geeks can be a little strange, and they pronounce this as Gooey. It is the environment that we all are familiar with when we use a standard computer. In the last chapter, we discussed a few programming languages that used text-based commands that mimicked English. For PLCs, a special type of programming language was developed that makes it easy for industrial technicians to program PLC operations. This visual language is called Ladder Logic, and it's as gooey as you can get.

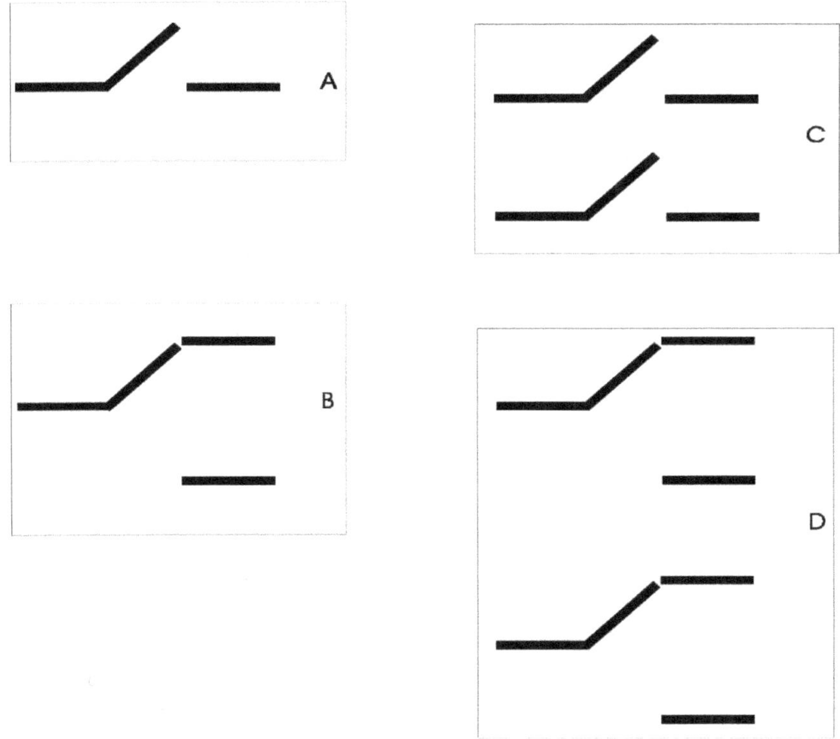

Ladder logic uses the symbology of contact devices to represent logic functions. Switches are the most common contact devices, and there are mainly four different contact types. Switch A in the drawing represents a Single Pole - Single Throw (SPST). It is the basic on/off switch that we discussed earlier in the text. A typical use for this type of switch is a standard light switch in the home. The switch depicted in drawing B is a Single Pole - Double Throw (SPDT). It's an on/off switch, but, with two possibilities. In the condition shown in the drawing, it would supply current to the top circuit, while the bottom circuit would be off. The condition would reverse in the opposite state. Typical use is as a two-way light switch in the home, where you can control a single light from two separate locations.

The switch shown in figure C is a Double Pole - Single Throw (DPST). This switch allows for the operation of two separate circuits and is basically like two SPST switches like that shown in figure A, which operate in unison with the top and bottom movable wipers ganged together. A typical use would be in machinery or electronic circuits. The last type of switch is shown in figure D and is a Double Pole - Double Throw (DPDT). It allows for two possibilities similar to figure B, but with the two poles operating in unison since the top and bottom movable wipers are ganged together. Again, a typical use would be in machinery or electronic circuits. The switches could be operated manually as in the light switch example or operated automatically as in a sensor device that may respond to temperature, pressure, or some other stimuli.

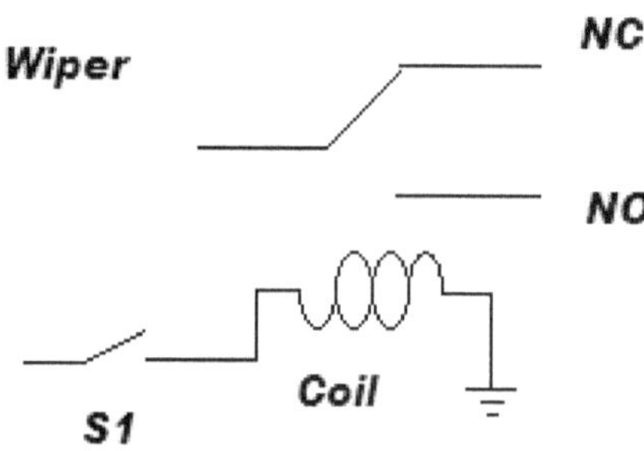

Figure 7.4. Electro Mechanical Relay

The other common contact device that is used in ladder logic is the electromechanical relay. The switch S1 shown in Figure 7.4 is an external control switch used to energize the coil inside of the relay. When the coil is energized and has current flowing through it, a magnetic field is produced that causes the mechanical wiper of the internal switch to make contact with the NO line. NC stands for *Normally Closed*, and NO stands for *Normally Open*. Relays are usually drawn in their de-energized state. The relay is a way to control the operation of remotely switched contacts through electrical means. The reason to use a relay might be that you want to operate the switching contacts from a distance, or perhaps you need isolation from the main circuit due to high voltage or high current in the switched circuit. A typical use would be in machinery or electronic circuits, and a very common application is in the starter circuit of an automobile. There is a massive amount of current that is required to operate the starter motor of a vehicle, and the large diameter wires that are connected to the starter would be difficult to run through the steering column. In the relay schematic shown in Figure 7.4, the switch S1 would be the car key, which can energize the relay coil. When the relay is energized, it supplies current to the starter motor through the NO switched contacts. In an automotive application, a relay is called a *solenoid*. Electromechanical relays were very popular but now are being replaced by solid-state relays that contain no moving parts. Many PLC programs use timers, counters, and other functions, but switches and relays are the most commonly used items when programming in ladder logic.

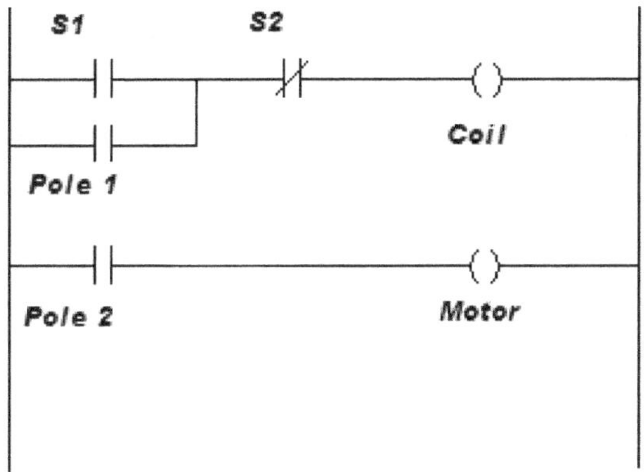

Figure 7.5 Ladder logic

Ladder logic is shown in the above Figure 7.5. The application of this program is to push and release start switch S1 to run a motor until stop switch S2 is pushed. In ladder logic, the vertical lines are a hot and neutral voltage line, and each current path is a rung on the ladder. In the drawing, switch S1 is a Normally Open (NO) switch symbol in ladder logic. When it is pushed, it makes contact to provide current to the internal relay coil of the PLC through the Normally Closed (NC) switch S2. The internal coil produces a magnetic field that causes the NO internal relay contact for pole 1 to close, thereby latching a current path even after NO switch S1 opens. The internal coil also closes the NO relay switch for pole 2, thereby providing a current path to run the motor. The motor will continue to run until the coil is de-energized by pushing the NC switch S2, which will open and stop the current through the internal coil, thereby releasing both internal NO relay contacts. All that is needed to make this a functional ladder logic program is to assign addresses to the various symbols that are used in the graphical representation. This type of visual programming procedure requires a programmer to follow current flows. It becomes much easier to program like this with practice, and an in-depth understanding of series and parallel circuit analysis, which we will study in the next chapter.

Section 7.5. Specialized programs

There are incredibly specialized computer programs available for applications such as RF antenna modeling, multi-layer printed circuit board layout, and static and dynamic force analysis of structures. Because of the limited number of users that might be interested in the purchase of programs such as these, the high development costs must be passed along to each purchaser. As it is with any product, as the number of items sold goes up, the fixed costs can be spread out to all of the consumers, and the cost per unit goes down. With specialized software programs, there is not as large of a market, and unit cost may be very high for each user. Extremely specialized programs may cost the end-user upwards of thousands of dollars per program license. We will introduce you to a couple of software packages that are specialized but widely used in engineering. The costs are more than a typical program, but not overly expensive. They may be in the range of hundreds of dollars, rather than thousands. The first program will look at is called *AutoCAD*.

Computer Aided Design (CAD) is the way that drafting is done today. Years ago, most companies involved with technology had drafting departments that would prepare engineering drawings using paper and pencils. The engineer would pass along a detailed sketch, and the draftsmen would formalize it as a print. Due to a consolidation of the workforce and the efficiencies generated because of the computer, many engineers are now required to draft a print using a CAD program like AutoCAD. This process is especially true in small companies with a limited workforce. There are several drafting programs available at a reasonable price, and AutoCAD is one of the more popular. After a CAD drawing is made, it can be rendered in 3D, and with additional software like Inventor, objects can be animated and interacted within *virtual reality*. These types of software packages are very intensive and require robust hardware. As it is with the latest computer games, modern versions of CAD software won't run very well on older PCs. The machine needs to be fast, have good video, and lots of RAM memory.

Even with the drawing made on a computer, you need to plan your work. A quick, simple sketch can dramatically aid in the development of a CAD project. There are two ways to represent an object. The first way is to look perpendicularly at each surface, where up to six normal views are possible. It is typical to show a minimum of three views, the front, side, and top. If it substantially aids in describing the object, additional views called auxiliary views could be drawn; otherwise, hidden lines could be used to show features not visible from the specific perspective. Solid lines would define visible edges, and hidden lines are drawn as broken lines. In constructing each of the views, you are describing them as a two-dimensional plane. This is called an *orthographic* projection because, in essence, you are projecting each feature of the object on to a parallel plane located a small perpendicular distance away. The system that is used for dimensions is Cartesian Coordinates where x =

length and y = height. This is also referred to as *rectangular coordinates*. Remember back in the first chapter, where we described our position in a room? The next step for realism is to move into three-dimensional space where x = length, y = width, and z = height. Rather than showing the sides of an object head-on, this second method of representation is called *Isometric* drawing and is more like a picture of the object as seen from the edges.

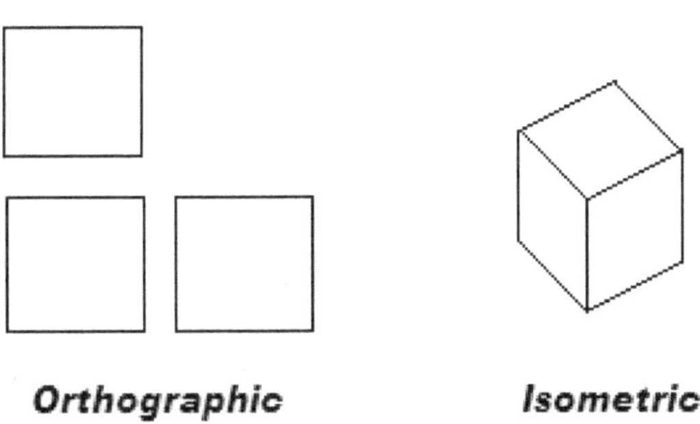

Just as in the first chapter where we mentioned that we sometimes wanted to describe our location in a room by a distance from the point of origin, you may sometimes wish to dimension CAD drawings in this way. This concept is especially true for curved surfaces. We learned about the Pythagorean Theorem in chapter one and used it to find direct distances. In AutoCAD, you can quickly convert to this type of dimensioning, showing the distance and angle, called *polar* coordinates. The AutoCAD program takes a little time to learn because it is a fully functional drafting program used extensively in industry. The user interface looks similar to the other programs that we explored. It's not too hard to start making simple drawings, and with a little practice, you can create very complex drawings for many applications in architecture, electronics, or mechanical engineering.

There is an even more specialized software package for electronics. It is produced by National Instruments and called Electronics Workbench or Multisim. The program allows the user to quickly drag and drop a wide variety of components to a work area, attach wires, and to connect virtual test equipment to the circuits. The amazing feature of this program is that it can simulate the operation of circuits with a great deal of accuracy. If the designer has an idea of a circuit, the program allows them to draw it and to test it, before building it on a workbench. A module of the program called Ultraboard can be used for circuit board layout after the schematic is designed. This suite of programs is widely used in schools and industry, and although more expensive than a general use program, the cost is reasonable. Eagle is another popular software package for circuits and for circuit board layout.

Many software companies have Web sites that allow downloading free trial versions of the latest software packages for evaluation. There may also be educational versions available at a reduced cost, but they may have slightly limited functionality. It seems that software companies have stepped up the time frame for program revisions and are releasing new versions of programs almost yearly. Earlier, we were talking about computer operating systems like Windows that act as an intermediary between an application and the computer hardware. The latest Microsoft OS has some very good updated features; however, there was some dissatisfaction with past offerings like Windows ME and Vista and Windows 8. Sometimes it is unnecessary to change to a new program or OS just because the software company makes a new version. Because of the need to keep overhead low, many businesses are running somewhat older versions of many software packages. There are even some companies that have low-tech applications still running with the old DOS operating system. That OS predates Windows. In the service industry, there is a very good saying, "If it ain't broken, don't fix it." As technologists, we always like to have the latest and greatest, but companies only spend money to make more money. Hopefully, not too many of them are still using DOS.

Chapter Seven Summary

Specialized computers may be embedded in machines like the Arduino microprocessor or generally designed to operate in a harsh manufacturing environment like the PLC. In both cases, the I/O ports will establish connections with input sensors and output actuators. The hardware need not be super-fast since things in the real world are not too speedy. The Arduino uses a version of C++, whereas PLCs use a visual language thought to lend itself to a manufacturing environment called ladder logic. Ladder logic is based on connections between vertical power busses, which look similar to the rungs on a ladder. Specialized programs are available for architectural drafting and electronic simulation and layout and usually require robust hardware.

Chapter 8
Series circuits

Section 8.1. The difference between electricity and electronics

A lamp in a room is an *electrical* device, while a television located next to it is considered to be an *electronic* unit. The lamp may only consist of a bulb and switch to give you light, but televisions are constructed of hundreds of electronic components working together in unison. It may seem that the interaction of the components in something as complicated as a television receiver could take years to understand, but nothing could be farther from the truth. Even the most complicated electronic products in our homes, such as televisions, game consoles, and computers, are constructed from just a handful of basic components. There are many slight variations, but the main electrical components that make up electronic products are Resistors, Capacitors, Inductors, and Semiconductor devices such as Diodes, Transistors, and Integrated Circuits. Just a handful of electrical parts, but their interaction produces the magic of modern-day electronics. It's just like the game of Chess, where there are only a few game pieces, but the knowledge of how the game pieces interact enables a player to master the game.

The electronic products we take for granted have been in existence for the equivalent of a blink of the eye through the long history of humankind. In the roaring 1920s, KDKA - the world's first commercial radio station went on air in Pittsburgh, Pennsylvania. Televisions became popular in American homes in the 1950s. IBM introduced the first PC in the early 1980s, and the Internet became a part of our lives in the 1990s. The study of electricity dates back much farther than the inception of radio, TV, and the Internet. It was the ancient Greeks who first studied the interactions of positive and negative charges. In American history, we have all heard the stories of Benjamin Franklin, who experimented with electricity by flying kites in thunderstorms.

Ben Franklin was able to explain electricity quite well for his day. From the first studies of electricity, early scientists knew an invisible electric force was at play in the universe. Benjamin Franklin helped explain the electric force and the current that it produces with a very intuitive approach. He related the electric force and the resulting flow of current to everyday situations. For example, if you hold something up high in the air and let it fall, it will fall due to the force of gravity, from a higher potential to a lower potential. If we examine the water current in a river, we understand that the gravitational force causes the water to run downhill. Common sense dictates to understand current flow in an electric circuit, that the flow will be from that of higher electrical potential to one that is lower. We sometimes still use Benjamin Franklin's common sense explanation of electric current flow today, and we call it conventional current flow. It is easy to understand, but it is backward!

It took a lot of scientific research to get from the early history of electricity to where we are at today. We now know that the movement of sub-atomic particles is what constitutes an electric current. A common-sense representation like the water flow in a river is still a great way to understand electric current flow, even though we now know that electric current actually flows uphill – from a negative to a less negative point. But if a gallon of water were to flow uphill or downhill, it's still a gallon of water flowing. Because it doesn't make any difference in the final evaluation, it is better to use our intuition and consider current flow from positive to negative, the way that Benjamin Franklin did. In the remainder of the book, even though we know the electrons bump through the wires from negative to positive, we will not speak of it again! If anyone doesn't like it, they can fly a kite. We will use easy to understand conventional current flow to examine both electricity and electronics.

Section 8.2. Defining series circuits

We have been mentioning current flow throughout the text but have yet to give a precise explanation of a circuit. The best way to think about current flow in a circuit is to picture it as a *loop*. Many wireline telephone repair technicians will refer to the term loop when describing a telephone circuit. For current to flow around a loop, it first must have a complete path to allow electrons to flow, and there must be a difference of *potential*. The difference of potential is a voltage from a source of energy, such as a battery.

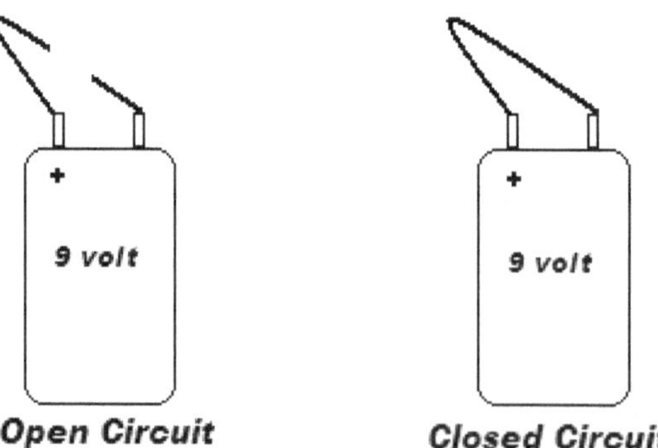

In the previous figure's open circuit loop, there is a break in the *conductivity* because the wire is not continuous. Electrons can move freely from atom to atom in a conductive wire, but the air has much less conductivity than wire and will offer a great deal of *resistance* to current flow. In an extreme case, such as during a thunderstorm, the movements of water droplets in the atmosphere cause a tremendous imbalance of electric charges. The voltage potential difference between charges becomes so high that it overcomes the air's resistance to current flow and produces a bolt of lightning. The energy release is so extreme that the heat of the flash is higher than the sun's surface temperature. Current can pass through the air at much lower voltages than what is needed to produce lightning. The voltage required to create a spark, sometimes also called an *arc*, can be reduced by moving the points of difference of potential closer together since there would be less air to resist the current flow. We haven't yet been able to harness the energy of lightning for a useful purpose, but we have many uses for arcs of current in our everyday lives. Some examples are the spark plugs in our automobiles, the igniters of gas stoves and outdoor grills, and electric welding equipment. However, arcs of current are usually a bad thing.

In our previous circuit diagram that connects a wire between the two terminals of a battery, we will have current, but unfortunately, too much current. The conductive wire forms a complete loop, but since there is no resistive element within the loop, it is called a *short circuit*. The wire won't get as hot as the air does with lightning passing through it. Still, the wire may get hot enough to melt or possibly cause a fire, and the battery could also explode because of the extreme amount of current flow generating heat inside the battery. Let's now take a look at Figure 8.2 with a current loop that won't blow up. The drawing on the right is a schematic representation of the circuit.

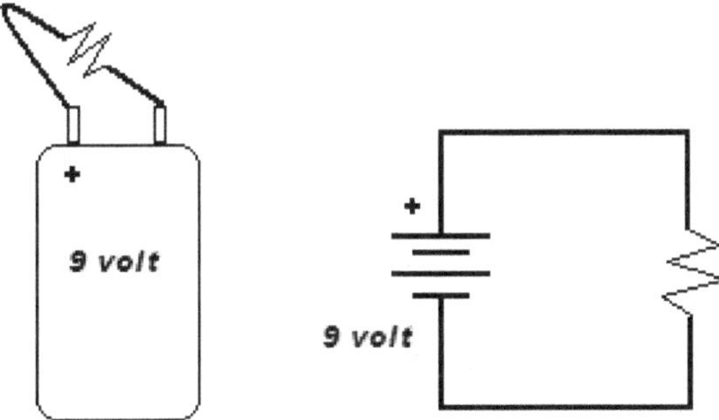

Figure 8.2. Resistive circuit

In our discussion of conventional current, we looked at a river's water flowing downhill. If you imagine a river that made a zigzag back-and-forth, it makes sense that there would be opposition to the water current. It wouldn't flow as freely as it would in a large straight channel because the zigzagging river would offer more resistance to the flow. Our latest circuit provides a loop for the current but limits its intensity of flow. The component that we added is called a *resistor*. The resistor's physical construction is such that it resists electron current flow, and the amount of resistance is expressed in the unit *Ohms*. As it is with the temperature of the air that lightning passes through, the resistor heats up as current goes through it, albeit nowhere near the temperature of the surface of the sun.

Along with the rating of the resistor's value in Ohms of resistance, there is also a rating in Watts that determines the maximum amount of heat that the resistor can safely dissipate to the surrounding air. The body of the resistor radiates heat. There are fins in a car's radiator to expand the surface area in contact with the air. Some large power resistors may have fins, but a typical resistor is cylindrical, and generally, resistors that are physically larger in size have a greater surface area to allow for heat dissipation and consequently have a higher rating in Watts. The value of the ratings in Ohms and Watts are totality independent of each other. The large resistor pictured in Figure 8.3 is an 8 Ohm Ω 20 Watt power resistor. The smaller size resistor is 11 Ohms Ω at 1 Watt. The horseshoe character Ω is the Greek letter Omega and is the symbol of resistance.

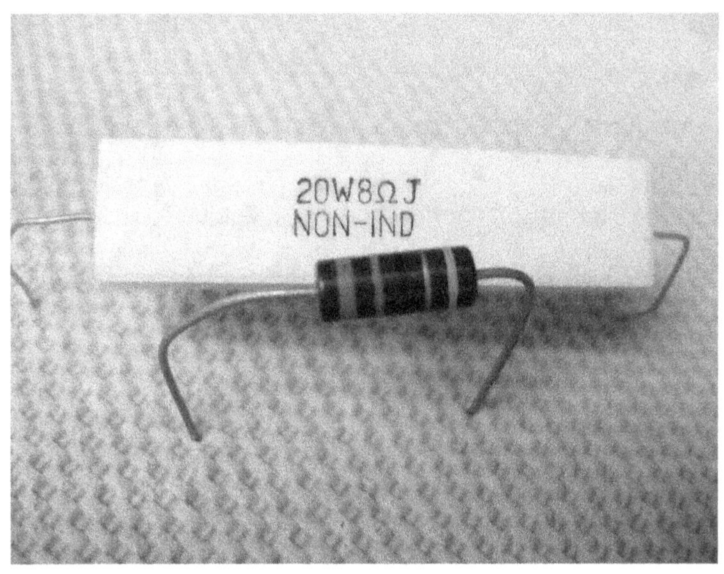

Figure 8.3 Resistors

A circuit with one loop like we have been discussing is called a *series circuit*. It's like if you were moving at a constant speed in a lunch line, each person would eventually get their food and pay the cashier. In an electrical series circuit, the current will all pass the same point, so the current amount is the same at any point in the loop. The formal definition of current is *charge in motion.* In math, it is explained as $I = \dfrac{Q}{T}$. Where I is the intensity of current, Q is the amount of charge, and T is the time in seconds. The unit of intensity of current is the *Ampere* (amp), and the unit of charge is the *Coulomb*, which is 6.25×10^{18} electrons.

<u>Problem</u>

A charge of 2 Coulombs passes a point in a series circuit in a minute. What is the amount of current?

<u>Solution</u>

Using the formula $I = \dfrac{Q}{T}$, where Q is 2, and T is 60 because there are 60 seconds in a minute:

$$I = \frac{Q}{T}$$

$$I = \frac{2}{60} = 0.033 \text{ Amp, or 33 mA}$$

(Remember that milli is the engineering prefix for thousandths.)

With one Amp of current flow equal to a Coulomb, which is 6.25 x 10^{18} electrons per second moving past a point in a circuit, the electron must not only be physically very small, but it must also have a correspondingly small amount of negative charge.

Problem

Find the amount of electrical charge for an individual electron.

Solution

Since 1 Q = 6.25 x 10^{18} electrons, if we call q the charge per electron:

$$q = \frac{1}{6.25x10^{18}}$$

$$q = 1.6x10^{-19} \text{ Coulomb}$$

The electron and Coulomb are not usually considered in practical applications, but Amperes are significant to consider in circuits. In most home consumer electronic items, the Ampere rating is usually in the 1 to 10 Amp range. If a fuse is blown or breaker tripped, there is usually a shorted or defective component that caused excessive current draw, and the unit will continue to blow fuses and oped breakers until a repair is made. Fuses and circuit breakers protect equipment from an over-current condition. Too much current could cause extreme damage to an electronic device or cause a house fire. When servicing equipment, never replace fuses or circuit breakers with one that has a larger rating.

Section 8.3. Series circuit analysis

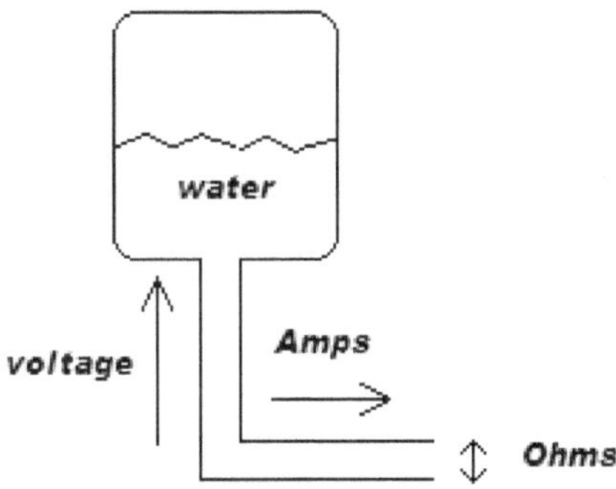

Figure 8.4. City water tower

We have used water flow as an analogy to electron current flow before. Let's think about it in that way one more time, before we begin our in-depth analysis of series circuits. If you imagine a city water system, a water tank would be high off the ground. The water is pumped up to the tank and stored there. It has potential energy due to the force of gravity. This equates to a battery, where through chemical action, a difference of potential exists across the terminals. In the city water tower, the main pipe diameter may restrict the amount of current flow if it is too narrow. This is analogous to a resistor in an electronic circuit. A thin electrical conductor would actually have more resistance than one with a larger diameter, and this is an important consideration for electricians who are selecting the gauge size for wiring. In resistors, the opposition to current is because they are typically made of a low-conductive material to resist current flow. As we saw in the last section, the component's physical size is not related to the resistance value. Since we now have a pretty good understanding of electrical qualities, we can develop a way to solve for unknown circuit parameter quantities. In a circuit, the three parameters that we will discuss are *voltage, current,* and *resistance.* From our previous discussion, we can see that the voltage directly relates to current flow, while the resistance is inversely related to the flow of current. This brings us to Ohm's Law. The formula can be manipulated to solve for any one of the unknown quantities.

For current: $$I = \frac{V}{R}$$

For resistance: $$R = \frac{V}{I}$$

For voltage: $$V = I\,R$$

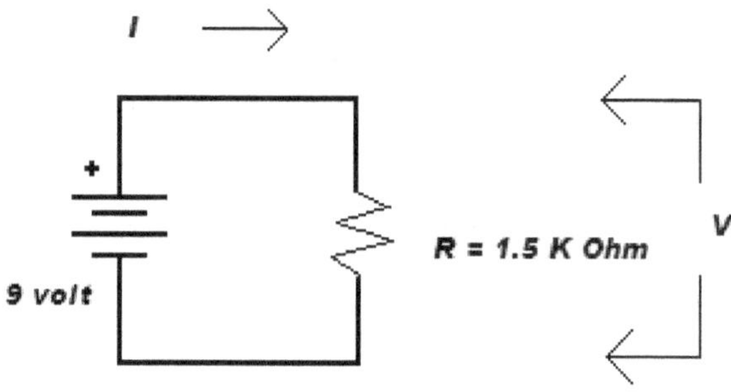

Figure 8.5 Circuit analysis

To analyze circuits, we need to understand the test equipment. A *Digital Multi-Meter* (DMM), shown in Figure 8.6, could be used to read the voltage, current, and resistance in a series circuit. To test for voltage, you would connect the test probes across the resistor, with the positive (red) toward the battery's positive terminal.

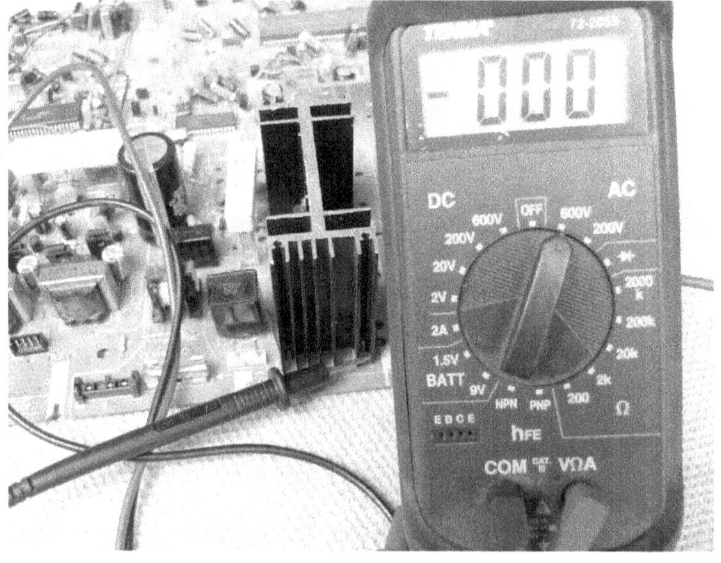

Figure 8.6. Digital Multi Meter

In our diagram with just one resistor, you would read the voltage of the battery. It doesn't matter if you connect to the wire near or far from the resistor because there will be no measurable voltage difference of potential along the wire. To use the DMM to test for current, you would need to break the loop at any point, and then insert the meter. The current will flow through the meter to measure it. In practical applications, testing current is sometimes impractical. To test resistance with a DMM, you must disconnect the battery and read across the resistor. You sometimes have to de-solder the resistor to not read other components connected to the circuit in parallel.

<u>Problem</u>

In Figure 8.5, use Ohm's Law to find the current in the loop.

<u>Solution</u>

$$I = \frac{V}{R}$$

$$I = \frac{9}{1500} = 6 \times 10^{-3} \text{ Amps, or 6 mA.}$$

We used the value 1500 Ohms because the k on the schematic is for kilo, which is the engineering prefix for thousands. In our answer, 0.006 Amps is 6 milliAmps, because milli is the engineering prefix for thousandths. Remember the chart in chapter one?

We also know that a resistor may get warm, as current flows through it. We will now introduce you to the power formula, sometimes called Watts Law. Like Ohm's Law, it can be manipulated and has three main versions:

$$P = IV$$

$$P = I^2 R$$

$$P = \frac{V^2}{R}$$

<u>Problem</u>

In the previous problem, find the wattage that the resistor would dissipate as heat.

<u>Solution</u>

Since we had to calculate the current I, there would be less of a chance for error using the formula that does not require I.

$$P = \frac{V^2}{R}$$

$$P = \frac{9^2}{1500} = \frac{81}{1500} = 0.054 \text{ Watts, or 54 milliWatts}$$

MilliWatts are thousandths of a Watt, which is very little power. That is about the power that many LEDs dissipate. Electronic communication hobbyists enjoy conducting low power contacts over great distances, and they sometimes make transoceanic communication contacts with transmissions of radio frequency power in this range. We will examine RF communications later in this text.

We will end our exploration of series circuits with a circuit of more than one resistor in the loop, sometimes called a voltage divider. In physics, there is a theorem that states conservation of mass and energy, saying that they can't just appear out of nowhere or disappear, but they can be changed from one form to another. An example is the transformation of chemical energy into electrical energy in a battery. There is also conservation in a circuit explained by *Kirchhoff's Voltage Law,* which basically says that the voltages across each resistor in a series circuit must add up to the battery's voltage. To solve problems with multiple resistors in series, you can use Ohm's Laws or a formula that combines two Ohm's Law formulas called the *voltage divider.*

$$I = \frac{V}{R} \text{ and } V = IR$$

Using the first formula to replace the I term in the second we have,

$$v = \frac{VR_x}{R_t}$$

Where: v is the voltage across a specific resistor, V is the applied voltage of the battery, R_x is the specific resistor's value, and R_t is the total resistance.

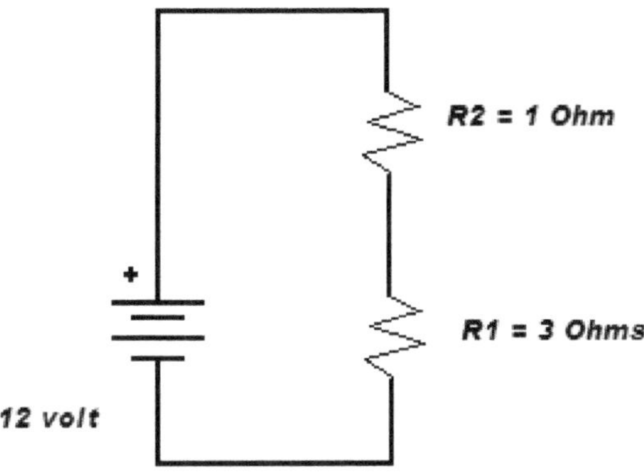

Figure 8.7 Series voltage divider

<u>Problem</u>

Find the voltage across each resistor in the figure.

<u>Solution</u>

Using the voltage divider formula for R1 we have:

$$V_1 = \frac{VR_1}{R_t}$$

$$V_1 = \frac{(12)(3)}{4} = \frac{36}{4} = 9 \text{ volts}$$

We could use Kirchhoff's Law and say that the remaining 3 volts must be across R2, or we can solve for it with the voltage divider formula.

$$V_2 = \frac{VR_2}{R_t}$$

$$V_2 = \frac{(12)(1)}{4} = 3 \text{ volts}$$

You may also notice that there is a ratio involved. We mentioned earlier in the text that proportions are a potent math tool. If we had a limited understanding of electronics, we could use pure math to find the answer to our problem using a proportion.

$$\frac{v_1}{12} = \frac{3}{4}$$

We are saying that the unknown voltage is to the total voltage of 12, as the 3 Ohms of the resistor is to the total Ohms of 4. We get the same answer with a proportion that we got using the voltage divider formula.

$$v_1 = \frac{(12)(3)}{4} = \frac{36}{4} = 9 \text{ volts}$$

Chapter Eight Summary

Electrons flow in a current loop from negative potential toward a less negative or positive voltage potential. We analyze circuits using the conventional current flow concept because it makes more sense and gives the same result. Most Digital Multimeters are capable of measuring resistance, voltage, and current. It is always necessary to measure resistance in a circuit with no power applied. Sometimes when measuring resistance in a circuit, one lead of the resistor may need to be unsoldered from the board to get an accurate reading. Current is the same throughout a series circuit, but multiple resistances within the loop may drop differing amounts of voltage. The total voltage is equal to the sum of the voltage drops and is known as Kirchhoff's Voltage Law.

Chapter 9

Parallel circuits

Section 9.1. Defining parallel circuits

In our discussion of series circuits using the analogy of water, we stopped at the main line from the water tower because the water pipes in homes are parallel. If water went through a two-story house in series, the upstairs toilet would empty through the kitchen faucet, and the sink would empty into the washtubs in the basement. It would be tough to get your dishes or clothes clean. When the water enters your house, the water flow splits up and takes different paths. These other current loops are why we use the term parallel. Physically the lines do not need to be in a parallel orientation, but they all need to be separate loops.

In an ordinary flashlight, the cells are placed in series. In a similar way that the voltage drops across individual resistors acted in a series voltage divider, the series cells in a flashlight add together to give you the total voltage. If you place three D-cells in a flashlight with each cell at 1.5 volts, there is a total of 4.5 volts across the bulb. Cars have 12 volt electrical systems, and you would never want to put two car batteries in series since it would give you 24 volts, but sometimes they are placed in parallel. Some emergency vehicles such as police cars and ambulances have two batteries in parallel in regions with cold-weather climates. Sometimes car owners with mammoth car stereo amplifiers also do this to get more power. Woofers can use a tremendous amount of power. Having two 12 volt batteries parallel only gives 12 volts, but the amps can go higher than a single battery because the overall current capacity is additive. All of the electric outlets in your home are also wired in parallel.

We discussed *parallel* processing when we looked at how to make computers work more efficiently. We said that with parallel processing in a computer, a large number of bits were all processed simultaneously. Since the ones and zeros are represented by ranges of high and low voltage levels, it requires a single line for each bit. There are ways in communications electronics to send more than one piece of information at any one time. An example that everyone has heard is called *multiplexing*. You hear it every time you listen to an FM radio and have the stereo separation of the left and right channels. You could design a computer to carry more than one bit of data on a wire at a given time, but it would involve encoding and then decoding the individual data bits. This would require additional circuitry on both ends. Over short distances, it's probably better just to have a wire for each bit. It makes sense that parallel processing is very efficient since it's like having many lanes on a highway to carry the traffic. More cars can travel along the road this way. The problem with driving on super-highways in big cities is that it can get a little confusing. The electronic circuits that are parallel contain more than one loop for current. Each separate loop is called a *branch*, and the total current will divide between the branches. The usual way to analyze parallel circuits is to simplify them.

Section 9.2. Parallel circuit analysis

There are three possibilities in analyzing parallel circuits: there may be two or more paths with equal value resistors in each path, there might only be two possible paths with unequal resistors, or none of the above. As it is with a multiple-choice test, you will most probably be correct picking the answer, "none of the above." But we will start our analysis of parallel circuits with the special cases because they require less work.

In series circuits, voltages and resistances simply add together to give us the total. In parallel circuits, however, we must look at things a little differently. We already discussed voltages in parallel. The voltage across parallel batteries is equal to the highest battery's voltage. It's probably not a good idea to connect batteries of different voltages together in parallel since the larger one will try to charge the smaller one. It's OK to connect resistors in parallel, but the total resistance will always be less than the smallest resistor.

To find the total resistance of resistors in parallel, we will examine the special cases that we described. First, there is the special case where equal value resistors are connected in parallel. In that case, the following relationship exists:

$$R_t = \frac{R}{n}$$

Where R_t is the total resistance across the parallel resistors, R is the value of the resistor, and n is the number of resistors placed in parallel. We are using Figure 9.1 for our parallel example.

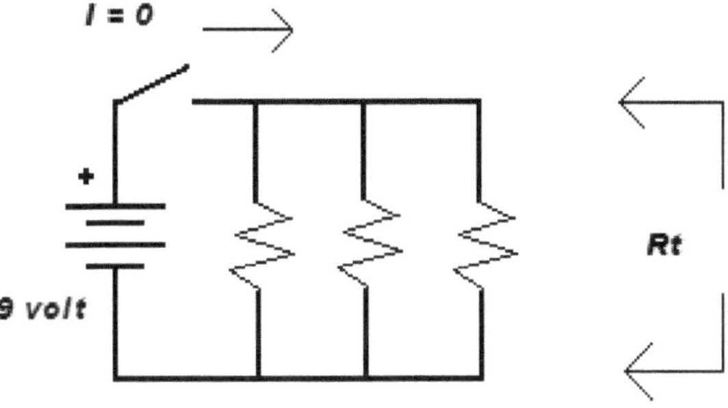

Figure 9.1 Parallel circuit

<u>Problem</u>

With three 100 ohm resistors connected in parallel, as shown in the schematic, find the total resistance value.

<u>Solution</u>

Because the resistor values are all equal, we can use the special case outlined above.

$$R_t = \frac{R}{n}$$

$$R_t = \frac{100}{3} = 33 \text{ Ohms}$$

We have the power source disconnected, so we could easily test the circuit for total resistance with a DMM. As we mentioned earlier, the total resistance of any number of resistors in parallel is less than the value of the smallest resistor value. Taking a moment to think about the voltage, if the SPST switch were closed, there would be 9 volts across each resistor, and you could use Ohm's Law to find the value of the branch currents. You really only need to find one, because they are all equal in our problem. The total current in the circuit would be the total of all the branch currents. The total current is in the main lines, before the branches separate it. This is called Kirchhoff's current law. If you ever have a problem analyzing an electrical circuit, compare it to how a city water system works. The current flow in an electrical circuit and water system flow work very similarly. In an electric circuit, the pressure is the force from a power source, the wires and components are like water pipes shown in Figure 9.2, and the electrons are like the water molecules.

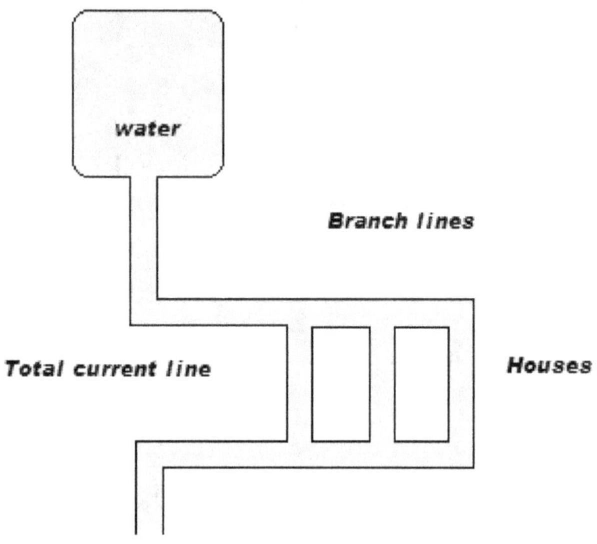

Figure 9.2 Water system

In our next special case of a parallel circuit, as in Figure 9.3, there are only two paths with unequal resistors. We can use the *product over sum formula*.

$$R_t = \frac{(R_1)(R_2)}{R_1 + R_2}$$

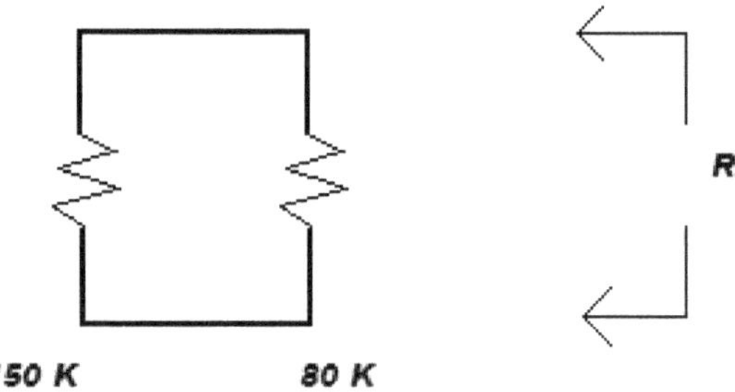

Figure 9.3 Two unequal resistors in parallel

110

<u>Problem</u>

Two resistors are connected in parallel. One resistor is 150 K Ohms, and the other is 80 K Ohms. Find the total resistance.

<u>Solution</u>

We can use the product over sum formula to solve the problem.

$$R_t = \frac{(R_1)(R_2)}{R_1 + R_2}$$

$$R_t = \frac{(150)(80)}{150 + 80} = \frac{12000}{230} = 52 \text{ K Ohms}$$

Since everything was in K Ohms, we just followed it through to the answer. If you are in doubt, however, you should enter the entire value in the calculator. You can use the numbers that we used but in thousands, or enter each number with the exponent 3. As a check for parallel problems, always make sure that the answer is less than the smallest resistor that is in parallel.

If we are asked to find each branch current in a parallel circuit like Figure 9.4, there are a number of ways to solve the problem. The most straightforward procedure is to use Ohm's Law, being sure to use the value of the resistance in each branch.

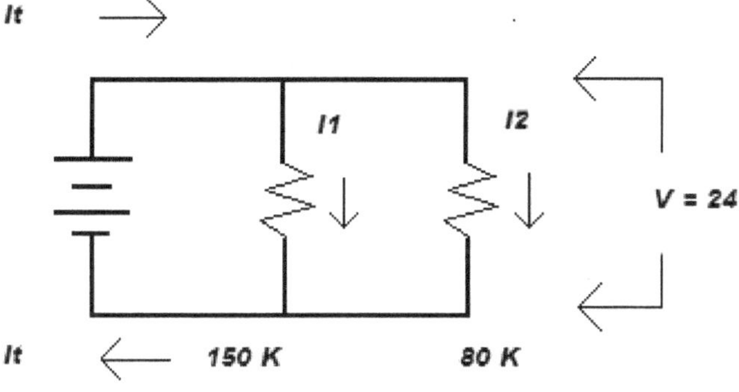

Figure 9.4. Parallel current problem

Using Ohm's Law for I1:

$$I = \frac{V}{R}$$

$$I_1 = \frac{V}{R_1}$$

$$I_1 = \frac{24}{150 \times 10^3} = 160 \times 10^{-6} \text{ Amps, or 160 uA}$$

The small u is the Greek letter mu and is the abbreviation for the engineering prefix micro. If you do not have your calculator in engineering notation mode, you would have calculated the answer as 1.6 exp^{-4}, which is the same result but expressed in scientific notation. Throughout the rest of this text, we will use engineering notation.

Next, using Ohm's Law for I_2 :

$$I = \frac{V}{R}$$

$$I_2 = \frac{V}{R_2}$$

$$I_2 = \frac{24}{80 \times 10^3} = 300 \times 10^{-6} \text{ Amps, or 300 uA}$$

Notice that more current goes through the smaller resistor. It makes sense that larger resistors will allow less current through because they resist it. If we wanted to find the total current I_t, we could simply add the two currents together. 160 uA + 300 uA = 460 uA. We could also calculate the total current using the total parallel resistance in the Ohm's Law formula.

$$I = \frac{V}{Rt}$$

$$I = \frac{24}{52 \times 10^3} = 460 \times 10^{-6} \text{ Amps, or 460 uA}$$

That is the same answer that we got when we added the two branch currents together, which we expected. You may have calculated a more precise answer, but remember back to an earlier discussion about significant digits? We have to round off because of inaccuracy in our beginning numbers. In reality, resistors commonly used in most electronic units have a tolerance of $\pm$ 5%, shown as a band of gold in the fourth color band. For a 150 K resistor, $\pm$ 5% is 7.5 K, either way of 150 K. The actual value could be as low as 142.5 K and as high as 157.5 K.

The last procedure that we will look at for solving parallel circuit problems is a very general approach. This method can be applied in all cases. It is a little tricky to enter into the calculator for the first few times, but we will explain the method step-by-step. The following formula is commonly called the *one-over* formula:

$$R_t = \cfrac{1}{\cfrac{1}{R_1} + \cfrac{1}{R_2} + \cfrac{1}{R_3} \cdots}$$

You might be comfortable using parentheses when entering a formula on a calculator. If not, then follow this procedure:

Step 1. Enter the value of R_1 and press the invert key. The key may be labeled x^{-1} or $\dfrac{1}{x}$.

Step 2. Press the + key, and then repeat step 1, but enter the value of R_2 and invert.

Step 3. Continue the procedure for all resistors in parallel.

Step 4. For the final result, press the = key, and invert the number by again pressing the x^{-1} or $\dfrac{1}{x}$ key.

<u>Problem</u>

Find the equivalent value of a 10, 20, 30, and 40 Ohm resistor in parallel, as shown.

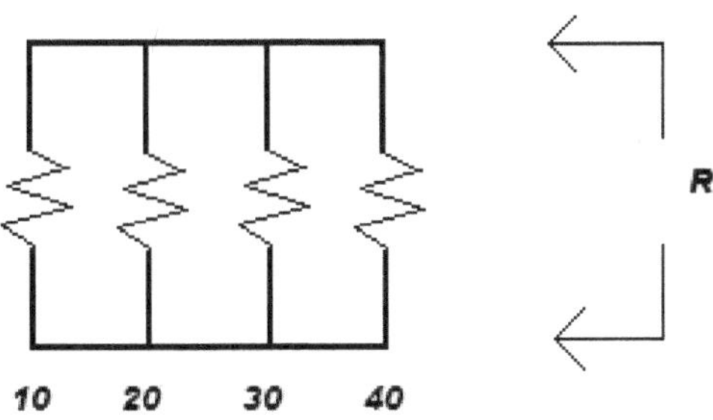

Figure 9.5 One over problem

$$R_t = \cfrac{1}{\cfrac{1}{R_1} + \cfrac{1}{R_2} + \cfrac{1}{R_3}\cdots}$$

$$R_t = \cfrac{1}{\cfrac{1}{10} + \cfrac{1}{20} + \cfrac{1}{30} + \cfrac{1}{40}} = \cfrac{1}{.208} = 4.8 \text{ Ohms}$$

Following the steps outlined in the explanation, we entered each value in the calculator, inverted it, and added all the values to get the denominator. Finally, inverting the denominator gave us the value of the total resistance. Notice that the total value is less than the smallest individual resistor.

A practical application of a parallel circuit is when people connect extra speakers to a stereo system. Most speakers have an impedance of 8 Ohms. Impedance is a complex number but can be thought of as the opposition to current flow similar to resistance. The problem some people run into when connecting extra speakers is that they go too low in Ohms across the amplifier output. As you add parallel branches, the equivalent total Ohms value decreases. This effect can cause an excessive current draw from the amplifier and may result in a blown speaker fuse, or damage to the amplifier. Speakers are also sometimes also damaged if too much current causes the speaker's voice coil to become over-heated and warp. As a speaker moves, it generates a reverse voltage called a back *Electro-Motive Force* (EMF) that keeps the amplifier from using excessive current and possibly becoming damaged. You can calculate the outcome of connecting parallel speakers with the formulas we have learned in this chapter. Sometimes people will add speakers in series and parallel so as not to lessen the Ohms to a dangerously low level. Series-parallel configurations can get a little complicated. With the solid foundation that we have now from understanding series and parallel circuits individually, putting them together won't be too difficult.

Chapter Nine Summary

Parallel circuits have branch currents. Adding the branch currents gives us the total current supplying the entire circuit. There special case formulas that can be used if applicable, or the general case where the one over formula can always be used. The total resistance in parallel is always less than the smallest resistor value.

Chapter 10
Complicated circuits

Section 10.1. Defining series-parallel circuits

If you look at the underside of a circuit board, you'll see many printed wire traces connecting all of the solder pads together. On most boards, there is a coat of insulation applied over the copper runs that may have a green or blue color. A long time ago, the boards were called Printed Wire Boards (PWB), but now they are called Printed Circuit Boards (PCB). Home consumer electronic units usually use a PCB printed on one side. Some smaller items may use boards printed on both sides. In computers, multi-layer boards are used with double-sided connections and many layers may be sandwiched together in between, to allow for the myriad of busses and connections between components.

Throughout the text, we have had a number of schematic diagrams of individual circuits. Functional circuits are rarely just series or parallel. As you connect circuits together, they form current loops that are both series and parallel. Looking at a schematic diagram of a complicated circuit is like looking at a map of interstate routes around the country. Finding your way through a schematic diagram of complex circuitry takes patience and a step-by-step approach.

Series Parallel Circuit

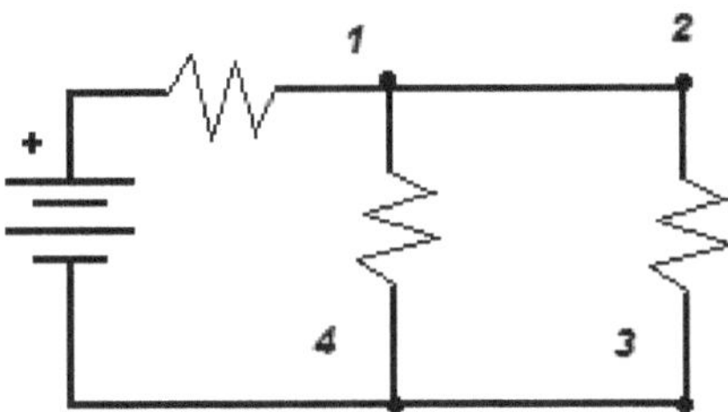

To navigate our way through a series-parallel circuit, follow current flow starting at the positive terminal of the battery. All the current must go through the first resistor in the figure because it is in series with the others. When we reach the *node* at point 1, the current divides with some current flowing through the first branch between nodes 1 and 4, and the rest flowing through the second branch between nodes 2 and 3. As we learned in the last chapter, the amount of current flow in each branch is determined by the amount of resistance in each branch. When the current splits at node 1 and flows through the first parallel resistor, it will recombine with the other current at node 4, and go in series from there to the negative terminal of the battery. These types of problems really aren't too complicated if you handle them step-by-step.

Section 10.2. Series-parallel analysis

It is always best in Series-parallel analysis problems to handle the parallel section first. After that, it becomes a series circuit, and they fall right in line! In our example circuit, we will first solve for the total resistance, and then apply Ohm's Law and Watt's Law to solve for other values.

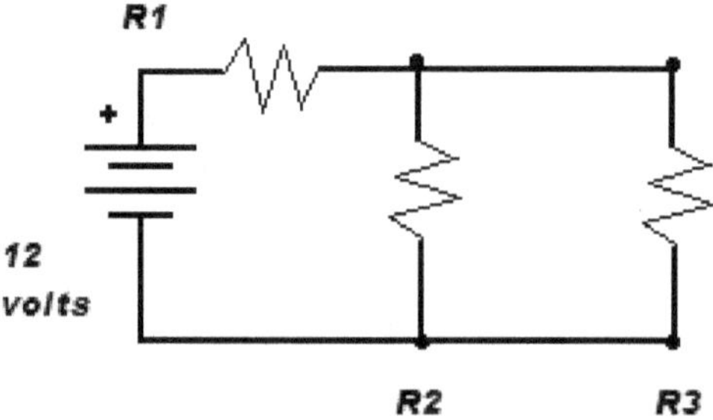

Figure 10.2 Series-parallel circuit

<u>Problem</u>

If all of the resistors in Figure 10.2 of the series-parallel circuit are 1 K Ohm, find the following values:

1. The equivalent resistance of the parallel section.
2. The total resistance of the series-parallel circuit.
3. The total current.
4. The current through each resistor.
5. The wattage dissipated by each resistor.
6. The total wattage of the circuit.
7. The colors of each resistor.

(Hint: remember to keep referring back to this the main schematic diagram.)

<u>Solution</u>

1. Using the product over sum formula in order to find the equivalent value of the parallel section gives:

$$R_t = \frac{(R_2)(R_3)}{R_2 + R_3}$$

$$R_t = \frac{(1000)(1000)}{1000 + 1000} = \frac{1,000,000}{2000} = 500 \text{ Ohms.}$$

It would have been easier to use the other special case formula in the last chapter since we have equal value resistors. In that formula, the resistor value was divided by the number of resistors, which in our problem is $\dfrac{1000}{2}$ = 500 Ohms. So, you can get the same answer in multiple ways.

2. Next, we will solve for the total resistance of the circuit.

In the previous step, we found that the parallel equivalent resistance was 500 Ohms, since we now have a series circuit you can just add the two values together as shown in the following Figure.

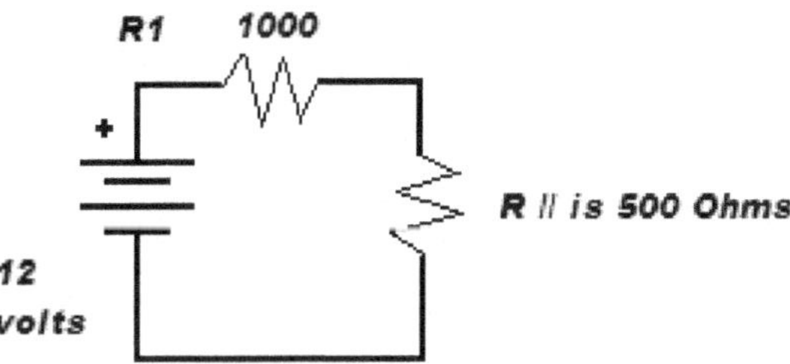

Figure 10.3 Finding resistance total

$$R_t = 1000 + 500 = 1500 \text{ or } 1.5 \text{ K Ohms}$$

3. Since we have total resistance, to find the total current, we will use Ohm's Law.

$$I = \frac{V}{R}$$

$$I = \frac{12}{1500} = 8 \times 10^{-3} \text{ Amps, or 8 mA}$$

It's not asked in the problem, but we could even find out how many electrons are moving through the circuit. One Ampere of current is one Coulomb of charge moving past a point in a wire in one second. One Coulomb is 6.25×10^{18} electrons. Our total current was 0.008 Amps, so the number of electrons is $(6.25 \times 10^{18})(8 \times 10^{-3})$. We'll hold it right there, that's a little too much detail!

4. Now we find the current through each resistor. The current through the series resistor is the total current of 8 mA. We are lucky in this problem because the equal value parallel resistors equally split up the current, but let's solve the problem in a more general way. This is the way that we would need to solve it with unequal parallel resistors. To find the individual currents through each parallel branch, we can use Ohm's Law for current, but first, we must know the voltage across the parallel section. We will use the voltage divider formula to solve for the voltage across the series equivalent resistance value of the parallel section.

$$V_{eq} = \frac{V R_{eq}}{R_t}$$

$$V_{eq} = \frac{(12)(500)}{1500} = \frac{6000}{1500} = 4 \text{ volts}$$

The answer makes sense from the relationship of the proportion. There is twice as much voltage dropped across the top resistor R1, because it is twice as big as the parallel equivalent value of 500 Ohms. So, 2/3 of 12 volts should be there, and 1/3 of 12 volts should be across the parallel section. This checks with our previous calculation.

Now we can continue to find the current through each resistor in the parallel section.

$$I_2 = \frac{V}{R_2}$$

$$I_2 = \frac{4}{1000} = 4 \times 10^{-3} \text{ Amps, or 4 mA}$$

The current for I_3 is the same.

$$I_3 = \frac{V}{R_3}$$

$$I_3 = \frac{4}{1000} = 4 \times 10^{-3} \text{ Amps, or 4 mA}$$

$I_2 + I_3$ = 4 mA + 4 mA = 8 mA, so it equals the total current, and we know that we handled both calculations OK. We could have also considered that because the values were equal in each of the two branches, the currents would have split equally.

5. Now *watt* else did they want us to find? Oh yes, the power dissipated in each resistor and in the total circuit. Remember the three power formulas that we looked at before? You can pretty much take your pick. We will use $P = I^2 R$. We need to remember to use the appropriate values for I and R.

$$P = I^2 R$$

$$P_1 = I_t^2 R_1$$

$$P_1 = (8 \times 10^{-3})^2 (1000) = (64 \times 10^{-6})(1000) = 64 \times 10^{-3} \text{ Watts, or 64 mW}$$

Next, we solve for the wattage of each parallel resistor, and we know that they are equal wattage because the currents and resistances are equal.

$$P_2 = P_3 = I^2 R$$

$$P_2 = P_3 = I_2^2 R_2$$

$$P_2 = P_3 - (4 \times 10^{-3})^2 (1000) = (16 \times 10^{-6})(1000) = 16 \times 10^{-3} \text{ Watts, or 16 mW.}$$

6. The total Wattage in the circuit is simply all of the wattage's added together, just as if you have three 100 Watt bulbs burning in a living room of a home, there would be a total of 300 Watts of power.

$$P_t = P_1 + P_2 + P_3$$

$$P_t = 64 \text{ mW} + 16 \text{ mW} + 16 \text{ mW} = 96 \text{ mW}$$

7. You may have thought that we were joking about finding the colors of the resistors. There actually is a color code for resistors, and it identifies the value in Ohms and the $\pm$ tolerance. The first color band identifies the first digit. The second color band is for is the second number. The third is the multiplier and usually tells you how many zeros to add after the numbers. Most resistors have a fourth color band that is gold, which means that the tolerance is $\pm$ 5%. The resistor colors used in our problem are all brown, black, red, and gold, as shown in the following chart. If you read the saying from top to bottom in the chart, the mnemonic is a way to help remember the number and color.

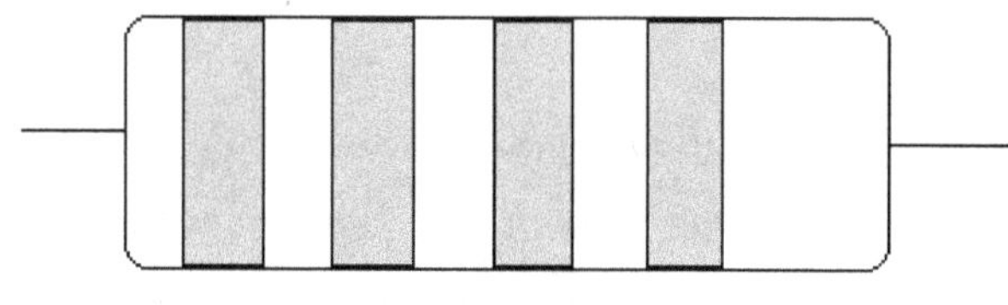

Number	Color	Mnemonic
0	Black	Bad
1	Brown	Bubble gum
2	Red	Rots
3	Orange	Our
4	Yellow	Young
5	Green	Guts
6	Blue	But
7	Violet	Veggies
8	Grey	Go
9	White	Well

Chapter Ten Summary

Series-parallel circuits can best be solved by collapsing it into a series circuit and working backward in a step-by-step process. Most real-life circuits are series-parallel. Resistors use a color code, and there are handy charts available to translate the code, or you can memorize a mnemonic.

Chapter 11
Noise and organization

Section 11.1. Entropy is elegant randomness

You may have heard the expression about trying to organize people as sometimes being similar to herding cats. People and cats tend to have individual plans and go their own way. But you really have experienced pure randomness, if you ever have heard the noise of static on your radio or have seen snow on unused TV channels. You may even have heard a cell phone make a noise on a computer sound system or some or stereo audio amplifier. The noise that the cell phone produces is interference, but it is organized interference, whereas the static on the radio or unused TV channels is disorganized noise. In general, the universe has a yin and yang duality between organization and disorganization. A great example of pure randomness is the behavior of plasma, the fourth state of matter. In plasma, the vast amount of energy causes atomic vibrations so intense that the material becomes dissociated, breaking atomic bonds and freeing the inner particles.

We all know about a substance like water that can exist in three states dependent on temperature. Water will be solid when the temperature is below that of the freezing point, a liquid while in a range near room temperature, and it will become a gas when the temperature exceeds the boiling point. But in all three states, there is an organization to the hydrogen and oxygen atoms and water molecules. But what happens when the temperature gets extremely hot? Just as things in the universe get strange as they approach absolute zero, odd things happen when we go the other way as well. We learned that heat causes the structure of matter to vibrate at the atomic level to an extent dependent on the energy level. At absolute zero, theoretically, there is no vibration - all motion stops. Above that, things are solidly frozen. As the temperature is increased, solid material such as ice will melt and take on a liquid state. As the temperature continues to increase, the atoms in the liquid gain enough energy to form a gas. As the temperature goes exceptionally high in regions of a star like our sun, the vibrational forces become so intense that a plasma disassociation may occur at the atomic level. At that energy level, the molecular bonds are broken, followed by the atomic bonds.

When the universe was born just after the big bang, the temperature was nearly infinitely hot, and there was almost complete randomness. As the universe expanded and the energy level diffused and became lower, matter was formed and gradually coalesced into an organized structure. Some believe that the universe will ultimately evolve at a point in the future into all matter with no energy. At that time, the universe will be dead, and there will be zero *entropy*. Entropy is the random motion of particles. A beautiful theory is that there will be enough force from gravity to begin pulling everything back together so the universe can restart with another big bang. At present, we have plenty of energy all around us, and it takes energy to produce interference, and we experience it from both organized and disorganized sources.

Section 11.2. Disorganized noises

One person's music may be another's noise. When we are calling something noise in this text, we are considering it to be interference. If it is noise from a disorganized generation source, it may be challenging to minimize its effects. An example of this type of interference is the microwave background emission that exists all around us. In the days of the Bell Telephone Company, scientists were looking for the source of noise that was in the microwave region of the electromagnetic spectrum. Signals in the range of 3 GHz to 300 GHz are considered to be microwaves. The interference was tracked using sensitive radio telescopes. The source was found to exist equally at every point in the sky. The scientists at Bell Labs found the remnants of the Big Bang - all around us. This constant noise comes from the very distant early universe, and it is a faint electromagnetic echo of an infinitely forceful event that occurred at the beginning of time. There are other continuous noise-producing sources in the cosmos and other sources that only produce noise periodically. The largest electromagnetic noise producer in space is our sun since it is near us in astronomical terms. Long-distance communication across our planet is greatly influenced by solar activity such as sunspots and flares. Not just communications are affected, occasionally some of the electrical power grids become overwhelmed by voltages spikes induced on the lines by the energetic extraterrestrial energy from a solar hiccup. The *magnetosphere* protects our world from the constant solar wind; it is a magnetic field that interacts with charged particles and sometimes treats us to auroras. The atmosphere also helps to shield our world from high-energy emissions produced in space. Ozone in the upper section of the atmosphere, called the stratosphere, protects us from receiving too much ultraviolet radiation from the sun. Ultraviolet light is above visible light, and exposure to it can contribute to skin cancer incidences by altering DNA. The energy of the sun also produces a thin layer at the edge of space called the *ionosphere*. It consists of charged particles that are beneficial to propagating long-distance high frequency (short-wave) communications. Thunderstorms in the lower atmosphere also contribute to the disorganized noise because of the electromagnetic radiation produced as lightning bolts discharge high electrical currents.

The random vibrations of particles at the atomic level that occur because of heat also are considered sources of disorganized noise. In an electrical conductor such as a wire, the random motion increases at the atomic level with increasing temperature. The thermal noise factor goes up with temperature because the conductor will have a higher resistance to organized current flow. (Mathematically, it has a positive temperature coefficient.)

In life, there are always tradeoffs. Higher heat may cause higher resistance in a circuit's conductors, but higher temperatures will serve to increase the chemical reaction in a battery. Part of why it is hard to start a car on a cold winter day is that the battery efficiency is reduced in cold temperatures. Some people take advantage of this relationship and store their batteries in a cool environment if they are not needed for an extended period. Semiconductors also have a more efficient operation as the temperatures increases. However, this attribute may be troubling in some circumstances, such as in transistors used for high power applications. As transistors conduct, they generate heat, making them perform more efficiently, causing even more heat. This paradox can lead to *thermal runaway* and the ultimate destruction of the device. Transistors used in these types of applications usually are located on heavy metal objects with fins to radiate excess heat, called *heat sinks*. Fans can also keep semiconductors running within a specified temperature range.

Section 11.2. Organized noises

Even if you were to reside on a deserted tropical island, you still would be subjected to organized human-made noise. There are high-powered transmissions that beam around the globe like over the horizon radar used for navigation purposes. Million Watt transmitters broadcast international radio programs on short-wave frequencies. People worldwide routinely rely on signals from satellites. Organized noise can also be produced by almost anything electric and radiate over great distances.

Figure 11.1. Neighborhood electric sub-station

On electric power lines, occasionally, the insulators between the wires and the poles can become contaminated and damaged and result in electrical arcing, which produces electromagnetic radiation. Even with the lines operating properly, electromagnetic *fields* exist near the lines. Studies exploring links between electric transmission lines and incidences of cancer have so far been inconclusive. Anytime current travels through a conductor, magnetic lines of force surround the conductor. For AC, the magnetic field is continually interchanging strength and direction with an electric field, and both propagate perpendicularly to the wire.

By their nature, electronic circuits are not only subject to interference but are also interference producers. Bipolar semiconductors are inherently noisy. As electrons jump over the PN junction, which is like an electronic speed bump, they must lose energy to go from a higher to a lower level. The electrons give off energy in the form of some combination of heat, light, and electromagnetic radiation. In an LED, the physical construction is designed to enhance the device's ability to emit light. When a semiconductor gives off energy as electromagnetic radiation, many problems with interference arise. The noise can leak into nearby circuits and cause disruptions. An electronic component called a *capacitor* can be used as a filter to eliminate some noise. A capacitor is constructed of two conductive plates separated by an insulating material in its most simple form. The effect is that an electrostatic charge is stored between the plates whenever a potential difference exists across the insulation. The stored charge can remain there for a time, even after the external difference of potential is removed. The device acts almost like a little battery. We

124

know that battery cells produce Direct Current (DC) from a steady, unwavering voltage. Capacitors like the one shown in Figure 11.2 are regularly are used in power supply sections of electronic units.

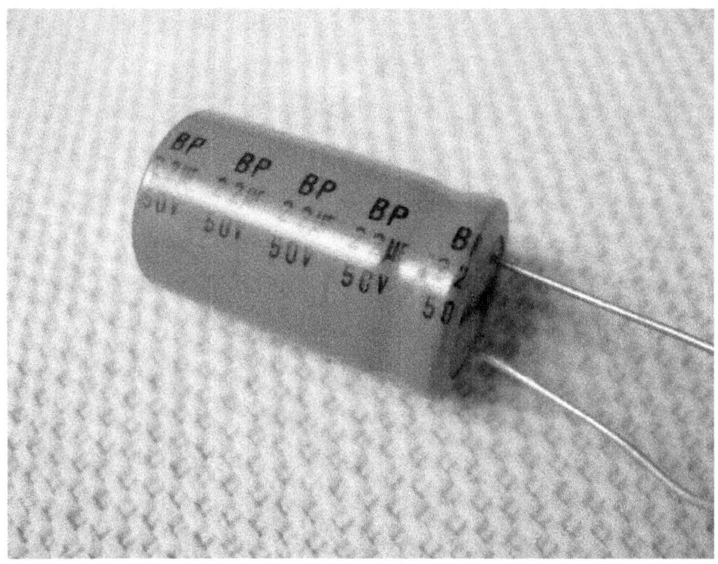

Figure 11.2. Electrolytic Capacitor

Since we have Alternating Current (AC) coming to our homes, power supplies found in almost every electronic unit must change AC into DC. This process is known as *rectification*. Positive pulsing DC is produced because the diode, shown with the arrow pointing to the right in Figure 11.3, acts as a one-way valve. A small capacitance across the diode helps suppress its noise, and a large capacitor across the power supply's output smooths the pulsing DC voltage.

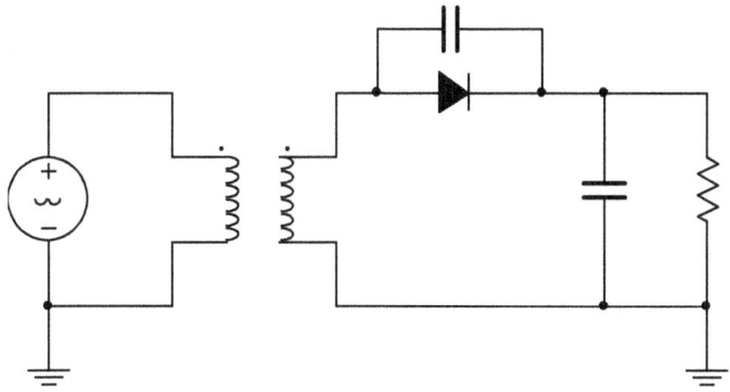

Figure 11.3. Power supply schematic

In our schematic shown in Figure 11.3, 120 VAC connects to the left's primary section of a *transformer*. This could be used for lowering the line voltage for charging cell phones and other devices. Transformers either isolate, step-up, or step-down voltages. In essence, they are two separate coils of wire that have an electromagnetic interaction that can allow changes in the voltage/current ratio from one side to the other. The diode is next on the secondary side of our power supply circuit. It acts as a one-way valve, only allowing the positive peaks of the AC alternations to pass. The pulsing half-waves keep the capacitor on the right charged up like a battery. The capacitor on the top of the circuit, across the diode, is used for noise suppression as the diode switches on and off. The capacitor eliminates the inherent switching noise of the semiconductor.

The noise of bipolar transistors is also problematic for creating interference. In our earlier example of a radio tuned to an empty channel, some of the background static is transistor noise. The noise produced by semiconductors is sometimes called *shot noise*. This effect can be particularly troubling when an exceptionally low-power signal is greatly amplified. Whenever you amplify a signal, the background noise is strengthened along with the information. This effect is a significant issue at the *front-end* of communication receivers, where signal levels coming from antennas are in the microvolt ranges. Silicon can be replaced with lower noise semiconductors such as Gallium Arsenide, and bipolar devices can be replaced with field-effect devices to lessen the shot noise. In satellite television broadcasts like Dish TV, a low noise amplifier is placed in the dish antenna's feed horn section. This keeps the tremendously tiny signals from encountering the resistance of the wire going to the house and minimizes the encounter with interference from sources along the way.

Motors, both industrial and those used by the average consumer, can produce massive amounts of noise called *Electro Magnetic Interference* (EMI). The worst perpetrator for the creation of local interference is the computer. The typical unit contains motors, semiconductor circuits, and a switch-mode power supply. The computer case is designed to shield much of the noise from leaking to the outside world, but it is not perfect. A great experiment is to place a portable AM radio, tuned to an unused channel, next to a computer as you progress through the boot-up sequence. You will hear many strange noises. If you try the experiment with the computer cover removed, you'll really hear some loud noises on the radio. That's part of the reason why computers should not be operated without covers in place for lengthy periods. It might also disturb proper airflow and can cause overheating.

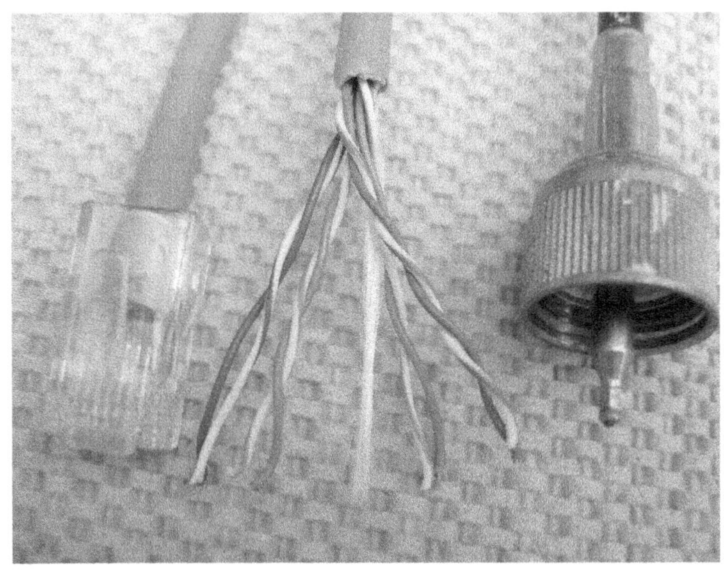

Figure 11.4. Cat-5, twisted-pair wire, and coax

Providing adequate shielding connected to an *Earth ground* is the best solution to resolving interference issues. A metal box surrounding an object allows currents to bypass whatever is in the box's interior. Physicists call this a *Faraday cage*. The metal jacket around a TV coax cable pictured in Figure 11.4, is an adaptation of a Faraday cage. The outer jacket is connected to Earth ground, and the RF signal is carried along the inner conductor. A very inexpensive way to reduce noise interference with ordinary wires is to tightly twist the two wires of a circuit loop around each other. Neither wire is connected to Earth ground, and the wires are said to be *balanced* or to *float*. Telephone wires use this type of conductor, which is called a twisted pair. It is also used in computer wires used for networking computers together. There are two twisted pairs of wires in a residential telephone cable, and 4 pairs in Cat-5 and Cat-6 computer networking cable. Some twin-lead television antenna ribbon cables used this same concept, but the conductors are held parallel to each other. Since the noise signals are 180 degrees out of phase across these unshielded wires, it cancels out.

Filtering is another solution used to solve interference issues, as we described in our power supply circuit. Along with capacitors, inductors can be utilized to suppress noise. Inductors are sometimes called chokes because they pass low frequencies but choke-off higher frequencies. Many times both capacitors and inductors are used in series-parallel combinations to suppress interference. In general, organized noise situations are somewhat controllable through proper design, filtering, and shielding. Equipment designers may use these steps to minimize interference problems. Controlling disorganized noise from external sources is a little less predictable, but using the same steps may reduce the effects.

Chapter Eleven Summary

Entropy is the name for random vibrations at the atomic level and increases with a system's energy. As the temperature of a conductive wire increases, the entropy increases and produces more resistance to organized current flow. Unwanted signals are considered noise and could be disorganized and random from sources like lightning or from cosmic origins. Organized human-made noises can come from current flowing through conductors, electronic devices, and machinery. Capacitors and inductors can be used to filter noise in circuits. Shielding, grounding, and twisting wires can also reduce interference problems in cables.

Chapter 12

Why Digital is Replacing Analog Technology

Section 12.1. The universe is analog, not digital

We have learned about circuits and some of their components. We looked deeply into the digital realm and know computers contain logic circuits, memory, and busses. We studied logic symbols, truth tables, and Boolean equations. But in nature, you would be hard-pressed to find anything that acted digitally. Digital is on or off, with no shades of gray. If sunrise were digital, it would be dark, but it would be light in the next instant. If the year's seasons were digital, it would be springtime, and you might need a jacket, but in the next moment, it would be summer and 80 degrees. The universe is analog!

The smooth continuous curved drawing on the left in the above figure represents a sound wave. The wave on the right represents a pseudo-analog output from a digital source, such as a cellphone. All sound is analog, but when sound is recorded digitally, the voltage value at each instant is assigned a binary numerical value and stored as digital data. The sound's conversion into digital information is done in a circuit called an *Analog to Digital Converter* (ADC). Each time an analog signal is converted to a number, it is called a *sample*. The faster the sound is sampled, the more accurate it will be. The size of the binary number assigned to represent the sound also matters. More significant bits equates to a more precise interpretation. On playback of the digital information, the conversion process again must take place. The conversion circuit for playback is called a *Digital to Analog Converter* (DAC). As it appears in the playback figure on the right, there will be a somewhat jagged response, which can be minimized through faster sampling and a binary number with more bits to increase accuracy.

It seems like a lot of trouble to go through when we need to convert back-and-forth between analog and digital. Still, there are so many advantages to having analog signals digitized that it is well worth the extra steps required for conversions. Digital media has the following benefits: Noise reduction, high-quality duplication, ease of manipulation and processing, and reliable transmission and storage. We will examine the advantage of using digital media over analog, one characteristic at a time.

Section 12.2. Noise reduction in digital circuits

Anyone over 50 years old may remember *vinyl records*. Some audio enthusiasts still use them, and they remain in limited production. Thomas Edison invented the technology, and aside from adapting the concept to hi-fi and stereo recordings, not much has changed. The vinal platters are recorded from an audio source used to vibrate a cutting needle. As the platter rotates at a constant speed, a spiral groove is made in a master copy. From the master platter, many copies are then pressed. When listeners play them on a record player turntable, the vibrations of a pick-up needle are converted into electrical signals which are amplified and sent to speakers. It is a very ingenious system; however, imperfections such as dust, dirt, mishandling, and even the playback needle slightly damaging grooves cause clicks and pops to interfere with the music reproduction.

Analog vinyl records became less prevalent in the 1980s, in favor of the *Digital Compact Disk* (CD), which uses a low-power laser beam for playback. The playback life of a recording on a CD is greatly extended since there is no direct contact with the disk's surface. Like its predecessor, the music is contained in a continuous spiral track, but optical reflections from the CD surface, or a lack of them, determines the sound. A *land* is a flat reflective surface, and a pit is a non-reflective point on a compact disk. To allow the data to be stored in a smaller geometric space, each actual bit is not directly associated with a reflective point. A code is employed to lessen the amount of stored data needed to reproduce the sounds. A CD is less than 5 inches, whereas album-length vinyl records are 12 inches in diameter. A good quality vinal record can reproduce high frequency sounds faithfully up to approximately 15 kiloHertz (kHz), whereas CD quality is up to 22 kHz. One Hertz is equal to one cycle per second, and humans can hear high frequencies up to 20 kHz

130

depending on the individual. As people age they tend to lose their upper-frequency range of hearing. Some retail businesses have been using high-frequency sounds as a teenage deterrent system. By generating a 20 kHz tone in their establishments, young people tend not to congregate because they find the tone annoying, while older people, who are the preferred customers, cannot hear it. Noise is all but nonexistent in music from a digital CD or any other digital form of recording. In the ADC when the music is recorded, the sampled sounds are assigned binary numbers, so when you hear the music that is produced by a digital source, you hear a representation of the music as it was recorded. Also, keep in mind that with digital information, there are no shades of gray. The bits are either on or off. The logic levels that we studied fell into a voltage range considered as a high or low level. We weren't concerned with the exact voltages because the range is all that matters for interpreting a logic level. TTL logic level ranges are shown in Figure 12.2.

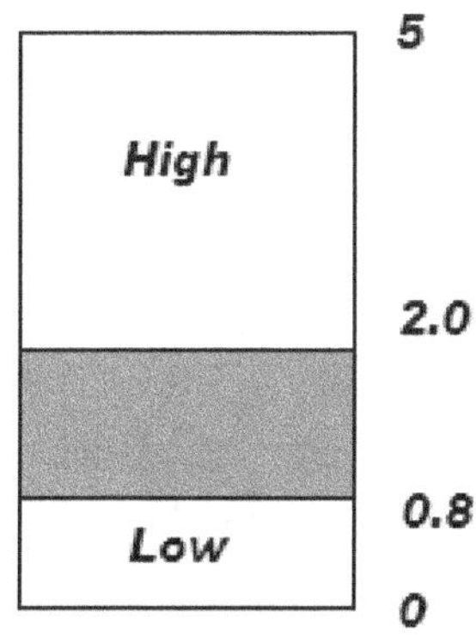

Figure 12.2 TTL logic levels

From our discussion in the last chapter, we know that electrical noise is all around us. It comes from nature and human sources, and the noise causes interference to our electronic circuits. In digital electronics, fundamental *noise immunity* is built into logic gates and other devices because only ranges are specified and not exact voltage quantities. Figure 12.2 shows that any voltage below 0.8 volts is considered a low, and any level above 2.0 volts is regarded as a high level. In the low output state, an IC is guaranteed to be no higher in voltage than 0.4. This allows for as much as 0.4 additional higher voltage of noise before going into the gray area. In the high output state, an IC is guaranteed to be no lower in voltage than 2.4. This allows for as much as 0.4 additional lower voltage of noise before falling into the gray area. As we said, voltages within the gray area are not permitted. Noise immunity of 0.4 volts, is allowing for a substantial amount of noise. In most cases, the interference would be far below that value. If equipment must function in an extremely noisy environment, another IC technology could be used, or noise reduction techniques could be employed. If the circuit were analog rather than digital, all noise values would be acted upon and passed along to the next circuit.

Section 12.3. High-quality digital duplication

Much to copyright holders' chagrin, it is effortless to make identical copies of digital information with no degradation of quality. The sales of CDs have dropped considerably with the advent of digital music sharing services available on the Internet. It is not only illegal but also unethical to make unauthorized duplications of the copyrighted material. When someone copies a song or movie that they have not purchased, it is akin to thievery. The problem is so prolific that only a small number of criminals are caught, but the prosecuted individuals usually pay a hefty fine. Agreements were made between content producers and computer software manufacturers to include a digital signature so that illegal copies would not function properly. It is certainly fair to make copies of your own work, and you can create as many copies as you would like. The final copy would be an identical version of the master copy, provided there is no compression using a lossy format.

Before the advent of digital technology, analog copying was limited because the quality would degrade with each copy. The degradation of quality for an analog recording can be noticed in the first duplication. Each time a copy is made, it is called a *generation*, and most analog copies are unusable by the fourth generation. The limitation is because every analog signal contains noise, and each successive generation will have all of the noise of the previous generations added to its own inherent noise. The noise can be heard as a hissing sound on audiotape and as a snowy picture on analog video recordings with a loss of picture synchronization.

Many popular Internet file formats such as Mp3 use *lossy compression* techniques to shrink the file size. There is a limitation on the number of times that these types of files can be copied. In lossy compression, an algorithm is used to throw out redundant information and eliminate material that may go unnoticed. Suppose you listen very carefully on a good stereo system to an identical song on a standard CD and the same song copied in the mp3 format. In that case, you should hear better quality from the song's original version. On the original digital version, nothing is lost in shrinking file size. The same compression issues exist with movies and pictures. The most popular compression technique for pictures on the Internet is

132

the JPEG format. Again, this format uses lossy compression to shrink the file size. After just a few generations, a noticeable loss of picture quality will occur. The picture may appear jagged and may have spots and blotches that appear in the photo that are called artifacts.

An uncompressed digital file can be copied for an indefinite number of generations. There are compression techniques for shrinking the size of digital files that are not lossy. The small file size is not as much of an issue today as it was in the past. Now we have larger storage devices, with most standard hard drives in the *terabyte* range. A byte is 8 bits of information, and tera is the engineering prefix for 1×10^{12}, which equates to a trillion. The number is slightly different in computers because of the base 2 system, but in any case, limited storage capacity is no longer as big of a problem as it once was. Compression over the Internet is still useful, but as we transition to high bandwidth connections such as fiber optics, the throughput is tremendous. As many computers go portable, and with HD broadcasting, bandwidth conservation remains a concern in over-the-air connections. HD digital TV uses very high compression techniques. Digitized high-quality video information is gargantuan in size and highly compressed to conserve space and reduce bandwidth. We believe that compression to reduce the file sizes of video will remain necessary for the foreseeable future.

Section 12.4. Ease of manipulation and processing

Before the computer, people used *typewriter* machines to do word processing. If the person typing made a mistake, they would have to use a quick-drying white solution or a small white strip of tape to hide the error before retyping the correct word. To save work meant making a paper copy and placing it in a file cabinet. How far we have come in just a few short years! Now information of all kinds can be manipulated in any number of ways through digital means.

Along with word processing programs used by just about everyone, there are many specialized programs for a variety of applications. In the field of art, graphic artists produce stunning works using the computer. Photographers once had to rely on manual airbrush techniques to retouch photographs. This can now be performed easily on the computer with a variety of photography programs such as PhotoShop. They can use such programs to enhance lighting conditions, sharpen camera focus, and modify any number of conditions that would be nearly impossible to vary without using digital technology. Medical diagnostic testing has advanced dramatically through the use of digital equipment. Perhaps the most impacted area of digital manipulation and processing is that of electronic communications. A few examples of tremendous advancements because of digital technology are computers, telephones, and television.

Many processes that were done using analog techniques can now be done more effectively through digital manipulation. Some compression processes are still done using analog methods. If, for example, we wanted to reduce the dynamic range of an analog signal to transmit in a small bandwidth but wanted it to appear unaltered to the listener, we could use an analog system called *companding*. In that system, changes in the volume of the analog signal are reduced by a certain amount. The reduced volume signal is then transmitted using less bandwidth. The receiver then boosts the changes in volume, so the listener hears the full *dynamic range*. Dynamic range is the change in the intensity from the softest to the loudest sound levels. In Frequency Modulated (FM) analog signals, the audio's loudness is directly proportional to the amount of electromagnetic spectrum needed to convey a given amount of information. A process like companding conserves bandwidth since it sends a low-volume signal. Because of the high degree of complexity, companding systems are in limited use, but another system is built into every FM broadcast radio. If you have ever had someone drive through your neighborhood with a loud car stereo, you might remember that you first heard the low-frequency bass rumble. The bass sound propagates better than higher frequencies. There is more power in the lower frequency sounds. The same situation exists for a radio station transmission, except they won't wake you at night unless you have your clock radio set to turn on. FM radio station transmitters boost the high frequencies sounds before they radiate the signal out of the antenna. This helps to keep the high-frequency sounds propagating as well as the lower frequencies. The boost at the transmitter is called *pre-emphasis*. The receiver has a circuit to reduce the high-frequency audio volume to bring it back to a normal level. The drawback of analog systems is that additional hardware is required to perform these tasks, and no changes can be made once the hardware is designed and installed. In digital signal processing, the engine is software-driven and can be changed quickly to accommodate different situations. For manipulating and processing digital data, all that is required is an algorithm to alter the data bits to accomplish the task.

Section 12.5. Reliable transmission and storage

In our previous examples, we illustrated a few ways to transmit and receive FM broadcasts using electromagnetic waves. There are many ways to convey information, and all that is needed is a transmission medium. You can use a method of communication that is directly perceivable such as communicating visually through sight like we are doing now on the pages you are reading. Sound can be used to convey information. Touch is another way for information to be directly conveyed. Direct methods utilize the five senses without conversion to some other form. The electromagnetic waves that we use for radio are useless to convey information unless they are converted to a form that our senses can understand. The indirect process of having a *carrier* that is varied in relation to a source of information to communicate over a distance is called *modulation*.

Frequency modulation (FM) varies the frequency of an electromagnetic radio carrier signal to transfer information. *Fiber optic* cable uses electromagnetic waves in the infrared or in the visible light range to carry information. The light can be modulated by analog means such as varying the intensity or digital means like flashing light on and off. There is very little chance of error over the fiber optic medium because the system is self-contained. Light can not enter or exit a fiber optic cable from the sides. There is more of a chance of errors occurring on an open channel like radio because of the medium's susceptibility to both organized and disorganized noise. A radio signal can be modulated both through analog and digital means. When some noise source produces interference with an analog signal, there is not much hope of overcoming the error. Digital signals can be repaired if corruption occurs.

Digital signals are usually exchanged in groups of bits called *packets*. A digital data communication system can skillfully be designed using a *parity* error-checking scheme. A complete packet can have a predetermined number of data bits with a single bit used for parity correction. The system can be designated as either an odd or even parity packet. The single parity bit will switch to conform to the correct parity scheme before transmission. The receiver will ensure that an odd number of ones are in the entire packet, or vice versa depending on the designation, or it will not accept the transmission and ask for a repeat. The parity system of error checking will fix one-bit errors, which mostly occur from a noise spike. There are even more sophisticated error correction schemes in use today that can be used to ensure the reliability of digital communication exchanges.

Storage for digital data can take many forms. Like it is with communication, a medium must be used to enable the exchange of information. In the storage of data, there are also direct and indirect methods. As with analog communication, if the information is stored in memory in analog form, there may not be a way to overcome an error. You can best safeguard data by making a backup copy whenever possible. There may be errors introduced in memory systems because of the storage mechanism's susceptibility to stray signals or faulty components. In short-term RAM memory, errors can occur and are dealt with by a parity or checksum process similar to the example just described. A computer hard disk drive is the most popular means to store long-term digital data. Standard hard drives use electric currents and magnetic fields to write and read data bits. In actuality, hard drives are shielded from stray fields quite well because of their grounded case. However, if nearby electric currents and corresponding magnetic fields were strong enough, the stored bits could be corrupted. On a mechanical hard disk, a small magnetic area's polarity on the platter is responsible for indicating if a bit is a one or a zero. For hard drives, typical methods use a mathematical process to both detect and correct errors.

Analog signals may be filtered so that damaged information is removed and not noticed, but the information that is missing will lessen the quality of the output. Digital transmission and storage methods are more reliable because error detection and correction methods can be employed. With error correction, the missing information may be reconstituted, and no information will be lost. Error detection and correction can be performed digitally through the use of complex mathematical algorithms.

Chapter Twelve summary

The universe is analog, but digital electronics has many advantages. Devices can be manufactured with simple circuitry in high density. Interference to the digital information can be negated through the use of voltage bands representing high and low logic-levels. Digital data signals are easy to process and store. Transmissions can be error checked and corrected.

Chapter 13

Analog systems and linear ICs

Section 13.1. Not everything can be digital

Digital systems use discrete quantities that represent the real world. If the real number line in mathematics were made of digital numbers, it would be a jumpy line since parts of the line would be missing. Some fractions have no discrete quantity. The number 1/4 is a discrete number, since when we convert it to a decimal, we see that it is the value 0.25. But, the number 1/3 is a repeating number, because when we convert it to a decimal, we see that it repeats indefinitely, 1/3 = 0.333... The three dots that follow the decimal number are called *ellipses* and indicate that the number continues. The number would need to be rounded-off to store it as a digital value. There are also *irrational* numbers like π that don't repeat, but go on indefinitely. We cannot give these kinds of numbers a discrete binary value without using an approximation, and that is why we say that the real number line would have a jumpy appearance if it were digitized.

Earlier in the text, we stated that the universe is analog. Other than at the extremely small *quantum* level, everything in the universe operates in a continuous manner. Using digital technology has many advantages, but it can only mimic real things. When digitization occurs, information is rounded-off to obtain a discrete value, some information is lost, and some error is induced. By keeping the step size very small, however, the error may be inconsequential.

There are some analog applications where discrete values are directly utilized. An example is a *stepper motor* used for accurate positioning in many robotic applications. The motor is analog in the sense that it has a continuous rotation, but it only stops at discrete positions of the rotation. Each position where the motor can stop is called a step, and stepper motors with a small step size can position to very accurate angles of rotation. Except for rare exceptions like this, most applications require a continuous response. When digital technology is used, it must almost always go through a two-way conversion process. The analog information first must be converted to digital data where it can be processed or stored, and then it must be converted back to analog for people to use it. We have already discussed the conversions used with the digitization of music. The ADC converts analog to digital.

After the data is processed or stored, the DAC converts the digital data back to analog.

Our ears respond to changes in air pressure. Greater displacement of the ear's diaphragm is equivalent to a greater volume. For us to hear a sound, a mechanical force must vibrate the air. The vibrations are in the form of acoustic waves. As sound vibrations occur, the air is compressed, resulting in a higher pressure as the object vibrates toward you. As the object's vibration moves away, the air is expanded, resulting in lower air pressure. These are called longitudinal waves. For one cycle of this process, the object must vibrate one time in both directions. The number of these cyclic repetitions per second is the frequency of the sound. Remember that one cycle per second is equivalent to one Hertz. The human hearing response is in the range of 20 Hertz to about 20 kHz. The air pressure vibrations move the diaphragms in our ears, and the faster the repetition of the movement, the higher is the frequency. The diaphragm is linked through chemical and electrical circuits to the brain, which interprets the signals as actual sound. Our speakers and microphones work in a very similar way.

Section 13.2 Transducers

A *transducer* is a device that transforms energy from one form to another. Similarly to the way that the ear works, a microphone responds to vibrations in air pressure. A microphone diaphragm is moved by the changing air pressure and converts the *acoustic energy* through mechanical energy into electric energy signals. There are many different conversion mechanisms employed in microphones. Old-style microphones used carbon particles that varied in resistance as the diaphragm either compressed or expanded their density. Some newer technologies use the compression and expansion of a crystal element to generate a voltage. Other methods include varying capacitance, inductance, or magnetic field. Once the transducer generates a voltage, electronic circuits can then amplify it.

When we use our voice, electrical impulses cause muscles to constrict and make the vocal cords vibrate. This transducer is an example that converts electrical energy into mechanical energy to produce an acoustic wave. A speaker or headphone is a very similar type of transducer. The speaker converts electrical energy into mechanical energy to cause the air to acoustically vibrate. Most speakers are electromechanical and use a movable coil of wire, called the voice coil, fastened to a heavy paper or plastic diaphragm. A current passes through the coiled wire to create a magnetic field. The magnetic field of the voice coil interacts with the magnetic field of a stationary magnet causing movement of the diaphragm. The polarity of the voltage across the coil, and the resultant direction of current flow, determine the direction of movement of the diaphragm. Most large speakers, such as woofers, work in this way. Tweeters and other small speakers tend to use a *crystal*. Just as the crystal microphone generates a voltage as the element is distorted by being bent back and forth by the moving diaphragm, a voltage signal sent to a crystal causes the crystal to move. A crystal will bend either back or forth, depending on the polarity of the voltage across it. Microphones and speakers can use the same concepts of operation in opposite ways. This is an example of *reciprocity*.

When we discuss electrical signals used to produce a sound, or generated from a received sound, we are dealing with signals representing waves. The waves are sinusoidal in nature. Remember that a sine wave has a positive and negative alternation. The characteristics of the sinusoidal sound waves are similar to the AC electric waves that we discussed earlier in the text. Electromagnetic waves are similar but contain an electric and magnetic component without any acoustic vibration of the air. The waves are analog because they do not have any *discontinuity*. When we studied digital electronics, we learned that transistors could be used to switch on-and-off to correlate to ones and zeros. In analog electronics, the transistor is operated between the completely on or off state, and used to amplify the analog signals that gradually change between high and low levels.

Section 13.3. Amplification

Most microphones and speaker systems are passive, which means they do not need electric power in addition to the signal for them to operate. An amplifier is active since a transistor needs applied power to be in the state between being completely on or off. We use the term *cut-off* when a transistor is completely off, and *saturation* describes a transistor completely in the on state. In digital circuits, the transistor is operated in these two extremes. The voltage across a *cut-off transistor* is maximum, whereas the current is minimum. The voltage across a *saturated transistor* is minimum, whereas the current is maximum, and vice versa. On a graph of current vs. voltage, a line drawn as in Figure 13.2, between the two extremes is called a *load line*.

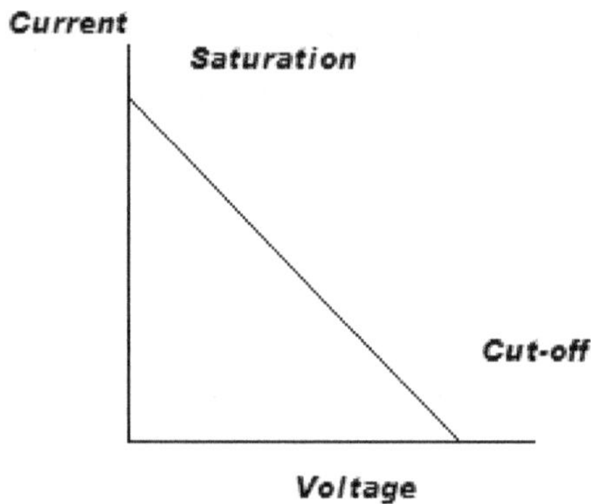

Figure 13.2. Load line

There are four main classes of amplifier operation relating to where conduction takes place on the load line without an input signal. That point is called the Q point and comes from the word *quiescent*. For the best fidelity, an amplifier's Q point is located in the middle of the load line. That location provides little chance of distortion from the signal driving the amplifier to either saturation or cut-off, causing a distortion condition called *clipping*. Clipping happens when the amplifier reaches its limit. This class of amplifier is considered *class-A*. This class has the best fidelity but poor power efficiency because it is conducting 100% of the time. The other classes conserve power but have a trade-off with the fidelity.

More than one transistor is usually needed to amplify small signals. Each transistor section will make the signal voltage greater by a certain amount called *gain*. We use the letter *A* to represent the level of amplification. Each section is called a stage, and the total voltage gain is the multiplication of each of the stages' gains. This is how many times the input signal voltage is multiplied by all stages of amplification, as shown in the figure.

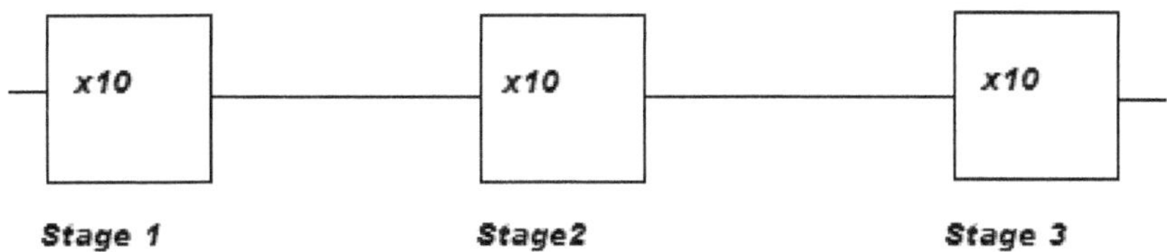

In our example, $A = (A_1)(A_2)(A_3) = (10)(10)(10) = 1000$. The input signal coming from the left is multiplied 1000 times as it goes through all of the stages and leaves on the right.

Problem

Find the final output voltage in the above figure if the input level is 18 mV.

Solution

The unit mV in the problem is in engineering notation, and the *m* is for milli.

$$A_t = (A_1)(A_2)(A_3) = (10)(10)(10) = 1000$$

The output $= (\text{Input})(A) = (18 \times 10^{-3} \text{ volts})(1 \times 10^3) = 18$ volts.

Notice that there is no unit associated with gain (A). Many times the gain of an amplifier is expressed in *decibels* (dB). The individual gains of each stage in dB are added together to find the total dB gain. For voltage, the following formula is used to convert to dB:

$$dB = 20 \log A$$

$$\text{Where A is } \frac{Output}{Input}$$

Problem

In the previous problem, express the individual gains and the total gain of all the stages using decibels.

Solution

For each stage:

$$dB = 20 \log A$$

$$dB = (20) \log 10$$

$$dB = (20)(1) = 20$$

For the total gain expressed in dB:

$$\text{Total } dB = 20 + 20 + 20 = 60 \text{ dB}$$

We could verify our answer by using the answer that we get for A_t directly.

$$dB = 20 \log A_t$$

$$dB = 20 \log 1000$$

$$dB = (20)(3) = 60$$

We have mainly been discussing voltage amplifiers. In order to drive speakers from a stereo system, the voltage amplifier stages must feed a power amplifier section. Class-A operation would waste a lot of power. Most power amplifiers are designed to run so that there is little current draw without an input signal. It is called push-pull operation and uses two transistors. One conducts during the positive going alternation of the input signal, while the other conducts during the negative half. The power amplifier Q point is very low on the load line diagram that we looked at previously, and close to cut-off. In problems that you wish to express power gain or loss in dB, the formula is slightly different:

$$dB = 10 \log A$$

There are a few quick rules of thumb for power expressed in dB. 3 dB is double power, and 10 dB is equal to a gain of 10 times. Sometimes there are circuits where you may want to reduce voltage or power. Reducing a signal is called *attenuation*. If a stage has less signal level coming out than is going in, the gain A and the dB will have a negative sign associated with it.

Section 13.4. Operational Amplifiers

With the rapid adoption of the IC, very few discrete transistor audio amplifiers are produced today. Entire analog circuits are contained in *linear* ICs. The word linear describes the operation. In linear amplification, the output is raised from the input signal in a direct relationship that is constant. If the input signal were plotted against the output, the graph would form a straight line. A versatile IC voltage amplifier called an Op-Amp is very popular because it is inexpensive, easy to use, and requires few external components. One drawback is that they sometimes require a split voltage supply to power the device. It must have equal but opposite voltages, one positive and one negative. There are, however, a few device types that only need a single supply. The Op-amp can be designed to either *invert* or *non-invert* the signal. Usually, nobody would ever know if the input signal was inverted as an output since it would sound the same. We will describe the inverting Op-amp operation shown in Figure 13.4 since it is easier to understand.

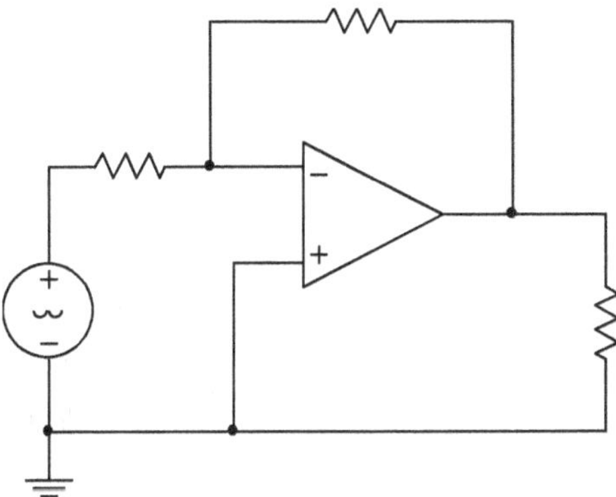

Figure 13.4. Amplifier Circuit

The diagram of an Op-amp is a triangle that points to the output. In our drawing a sine wave signal source on the left is connected through an input resistor to the negative sign on the Op-amp. The negative sign identifies the inverting input. The other resistor that is connected to that point is called the *feedback* resistor. It takes part of the inverted output signal and sends it back to the input. Since the output is inverted, it has a signal that is the opposite. The negative feedback subtracts from the input level. The gain of the stage is determined through the selection of the ratio of the feedback resistor to the input resistor. The other resistor on the right of Figure 13.4 develops the output voltage. The voltage gain of an inverting Op-amp can be found with the following formula:

$$A = \frac{R_f}{R_i}$$

<u>Problem</u>

Determine the gain of an Op-amp circuit if the feedback resistor is 10 kΩ and the input resistor is 1 KΩ.

<u>Solution</u>

Remember the symbol Ω is the Greek letter Omega and is the symbol for resistance.

$$A = \frac{R_f}{R_i}$$

$$A = \frac{10{,}000}{1000} = 10$$

The output voltage would be 10 times larger than the input.

Op-amps also need to have two IC pins connected to power the IC. They are not shown in our drawing. Using Op-amps in the design of amplifiers has many advantages. They have a high input impedance. You may remember that impedance is very similar to resistance. With a high input impedance, the stage in front of the Op-amp would not be affected by the connection. Op-amps also have a low output impedance. That is good because there is very little opposition in the Op-amps output to supplying current to the following stage, should it require it. Op-amps are designed to be voltage amplifiers. There are other ICs available that are power amplifiers. An Op-amp that is in wide use has the part number 741, and a common IC power amp is part number 386. For a linear IC, the prefix is a group of letters that identify the manufacturer. An LM741 is a 741 Op-amp made by National Semiconductor. There is a different system for digital IC part numbers. For digital ICs, the base number is preceded by letters that identify the type of technology, and the part number begins with 74 to identify the IC as commercial production, or 54 to signify that it meets military specifications. For a 74LS08, the 08 is the base number for an AND gate, LS is low power Schottky technology, and 74 identifies it as suitable for commercial use. It is very easy to search for an IC part number on the World Wide Web, and download a manufacturer's datasheet at no charge. The datasheet will list all of the specifications for the device, as well as a pin-out diagram to identify the function of each connection pin.

144

The pinout assignments for *digital* ICs are very standardized. On almost all digital Dual Inline Package (DIP) ICs, the ground connection is the lower rightmost pin, and the power connection is the upper leftmost. The reason for this standardization was because, in the past, many digital ICs needed to be connected together on a circuit board. This IC pinout orientation made it easier to lay out the ground and power busses on the board. Analog ICs do not adhere to this convention, and each device will have different pin assignments. The numbering system, however, is the same for all ICs. An orientation mark, usually a semicircle or circle, is placed on the left, as appears in Figure 13.3, and the lower leftmost pin is number 1. The numbers go along the pins counter-clockwise, with the highest number assigned to the upper leftmost pin.

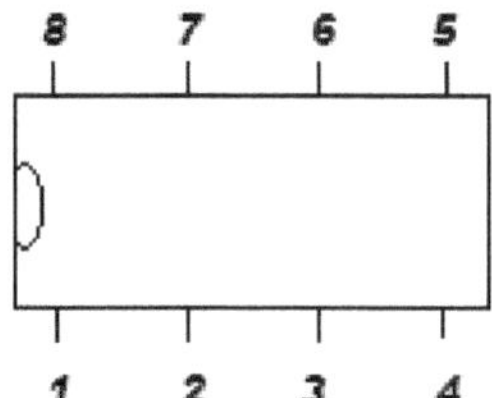

Figure 13.5. Pin numbering of an IC

Chapter Thirteen summary

Transducers are analog devices that convert energy from one form to another. Microphones and speakers are examples of transducers with reciprocal properties. Audio amplification is usually accomplished by using linear ICs like op-amps; however, with a little extra design effort, transistors can be used. Amplifiers are ranked by the amount of power they require and the output signal fidelity. Class A amplifiers are used in high fidelity applications for audio voltage amplification. Class B amplifiers (or AB) require less continuous power but can produce the amounts of currents necessary to power speakers. Both analog and digital ICs use the same pin numbering scheme, but you must check the specific datasheet for functional pinouts and IC parameters.

Chapter 14

Radio frequency communication

Section 14.1. There is more on radio frequencies than radio

If you're asked to name your favorite radio station, you might say Rock 104, Buzz 106, or Hot 101. The Hot-Buzz-Rock part is a positioning slogan, but the numbers are the frequencies of the radio stations. They are really an approximation of the frequency, rounded off to the nearest MegaHertz. The Federal Communications Commission (FCC) grants licenses to radio and TV stations for specific channels. On FM radio, the band of frequencies goes from 88 to 108 MHz. The stations are located on channels spaced 200 kHz apart, starting at 88.1 and ending at 107.9 MHz. All of the FM radio slots are odd in number. The band of frequencies starts at 540 kHz and runs to 1700 kHz on the AM radio dial. AM radio stations are positioned every 10 kHz because their bandwidth size is much less than FM. The bandwidth concept for radio is similar to the way it works on computers. A larger bandwidth can accommodate more information. That is why the wider bandwidth FM stations sound better than AM. Formally, this is called *Hartley's Law*.

$$\text{Information} \propto (\text{Bandwidth})(\text{Time})$$

This proportionality is easy to remember when applying it to computers. Suppose you are online and downloading a large file. With a given file size, greater bandwidth would mean less time and vice versa. The problem with large radio bandwidth is that there is a limited amount of electromagnetic (EM) spectrum. The concept is similar to commercial road frontage on a busy street, where every foot is valuable.

There is more than just broadcasting on the Radio Frequency (RF) spectrum. A few of the other services include police and fire two-way radio, cell phones, aircraft communication, and navigation. There are very few sections of the Electromagnetic (EM) spectrum that go unused. Below is a general idea of the classification of frequencies:

Low	Medium	High	VHF	UHF	Microwave
3Hz to 300 kHz	300 kHz to 3 MHz	3 MHz to 30 MHz	30 MHz to 300 MHz	300 MHz to 3 GHz	Above 3 GHz

Some of the frequencies that were chosen for the use of radio services by the FCC were arbitrarily made, while others are in certain sections due to bandwidth and *propagation* considerations. Radio waves propagate (travel) differently at different frequencies. The lowest frequencies are used for long-range navigation and specialized data communications. If you have an atomic clock, it receives signals from the government's National Institute of Standards located in Colorado, on a frequency of 60 kHz. The low frequencies tend to travel very far and follow the curvature and terrain of the Earth. Medium frequencies are referred to as Medium Wave (MW) and contain AM broadcasting transmissions and specialized communications. This range of frequencies is a medium-range for distance and normally doesn't propagate over 100 miles. There are exceptions, and on occasion, the radio waves can be reflected by the ionosphere and travel over great distances, usually at night. High Frequency (HF) is sometimes called short waves. This range of frequencies is very much affected by the ionosphere because most of the waves that go up into the sky are reflected down to the Earth. There are international broadcasts in this range, and Ham radio operators use the frequencies to make transcontinental exchanges of digital data and voice communications. Very High Frequencies (VHF), and above, are best for short-range communication. TV channels are in the VHF range, as well as FM broadcasting and many two-way radio services. UHF contains TV channels, as well as two-way radio services and cell phones. The frequencies above 3 GHz are classified as *microwaves* and are used for cell phones, Wi-Fi, and satellite communications. All microwave ovens operate on the frequency of 2.45 GHz. the devices must run on one frequency to reduce interference. Microwaves are small waves. As a frequency gets super high, it is easier to describe it in terms of a small wavelength. We discussed wavelength earlier in the text when we calculated the length of a cell phone antenna. The formula for finding wavelength is $\lambda = \dfrac{c}{f}$, where c is the speed of light and f is frequency. Satellite TV services operate in a frequency range upwards of 10 GHz. As frequencies go higher, wavelengths get smaller. Satellite TV wavelengths are very small, and parabolic dish antennas are used to focus the signal in order to increase the gain. Antennas are passive devices, but with the proper design of their physical construction, they can achieve a high gain by concentrating the signal.

Section 14.2. Modulation mixes two signals

It is debatable if the first method used to communicate on radio frequencies was modulation at all. Early in the Twentieth Century, the first professional use of radio was aboard ships. The devices were called a wireless and used Morse code like the wired telegraph did at the time. When the Titanic sank in 1912, the radio operator alerted the world by sending the signals " ... - - - ... ". Three dots, three dashes, followed again by three dots are the Morse code characters for S,O,S. The dots and dashes are similar to the binary communication that we use in our computers. The difference is that each letter, number, and punctuation mark has its own sequence of dots and dashes. In computers, we do something similar with *ASCII code* (pronounced ask-key). ASCII is an eight-bit code that represents all of the letters, numbers, and special characters on a computer keyboard. In the early days of radio to send Morse code, the operator used a telegraph key as pictured in Figure 14.1, to turn the transmitter on and off quickly. When it was on, it sent a continuous carrier wave (CW). Some Ham radio enthusiasts still use CW to this day. Modulation is defined as the process of varying one signal in response to another to convey information so that CW may qualify in the broadest sense.

Figure 14.1 Telegraph Key

It is not uncommon for CW operators to manually send upwards of 30 words per minute. However, even automated CW programs are slow by digital standards. Sending data pulses directly without modulation, as is done in a computer's busses, is called *baseband* communication. Pulsing transmissions with RF can produce interference. A wave transmission that is rapidly changed between two states produces *harmonic* interference to stations on other frequencies. A harmonic is a higher multiple of the fundamental frequency.

<u>Problem</u>

Find the first three harmonics of a signal at 100 kHz.

<u>Solution</u>

The first three multiples are 2, 3, and 4

$$F_1 = (2)(100 \text{ kHz}) = 200 \text{ kHz}$$

$$F_2 = (3)(100 \text{ kHz}) = 300 \text{ kHz}$$

$$F_3 = (4)(100 \text{ kHz}) = 400 \text{ kHz}$$

Square waves are made up of an infinite number of odd harmonics. As harmonics go farther away from the fundamental frequency, their power becomes less. Harmonics are called key-clicks in CW radio operation, and are minimized by slightly lengthening the rise and fall times of the RF pulse transmission as shown in Figure 14.2. Rise and fall times are measured between the 10% and 90% amplitude points. For digital logic signals, we want the levels to switch as quickly as possible with a very little rise and fall time. The sharp edges on square wave pulses help explain why computers generate a great deal of RF interference.

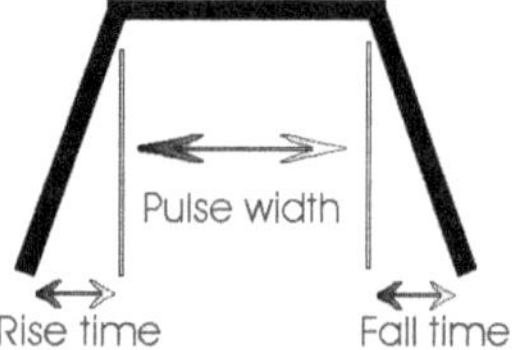

Figure 14.2 Avoiding Key Clicks

In amplitude modulation (AM), the carrier amplitude is varied in response to the audio signal. AM has been around a long time. Its beginnings date back to the same era as the Titanic. On one Christmas Eve, a radio operator played holiday music to the shipboard operators at sea using AM. That may be the beginning of broadcasting, as we know it. The modulated wave can be viewed by connecting an Oscilloscope to the AM transmitter's output.

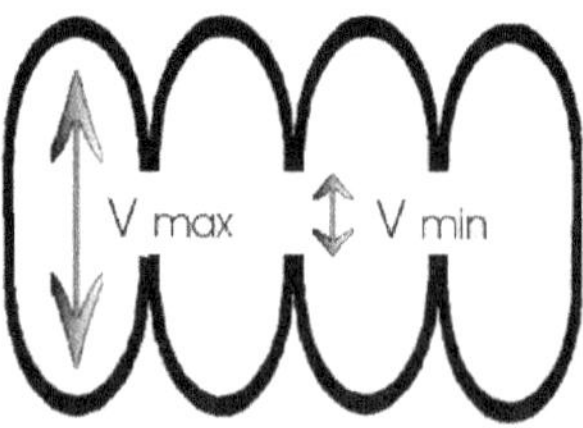

Figure 14.3 AM waveform

The loudness of the AM transmission is determined by the relative sizes of the changes in amplitude. The following formula can be used to find the *percentage of modulation*:

$$\% \text{ Mod} = \frac{V_{max} - V_{min}}{V_{max} + V_{min}} (100)$$

Problem

Find the percent of modulation of an AM radio signal that has a V max of 2 volts, and a V min of 1 volt.

Solution

Substituting the given values into the formula:

$$\% \text{ Mod} = \frac{V_{max} - V_{min}}{V_{max} + V_{min}} (100)$$

$$\% \text{ Mod} = \frac{2 - 1}{2 + 1} (100)$$

$$\% \text{ Mod} = \frac{1}{3} (100) = 33\%$$

On an AM signal, 100% modulation is full loudness, and 0% modulation is no sound at all. In our problem, 33% would not be very loud. Care must be taken to not exceed 100% modulation, or the signal will distort. The entire wave is called an *envelope*, and you can almost imagine the sound wave in Figure 14.3, riding on top of the carrier wave with its reflection on the bottom. The best way to find the percent of modulation is to use a single tone, and not music or voice. Music and voice signals are too complex to get an accurate reading on the test equipment. A common tone that is used is 1kHz for test purposes.

On AM, the envelope consists of three component waves, the carrier, the upper *sideband*, and the lower *sideband*. The sideband signals are transmitted above and below the carrier frequency by an amount equal to the frequency of the tone. An RF carrier can be thought of as a truck, and the intelligence signal is loaded on the truck for transportation. Figure 14.4 represents an AM transmission referenced to the frequency domain on the x-axis showing a carrier with the upper and lower sidebands of a transmitter broadcasting a single tone.

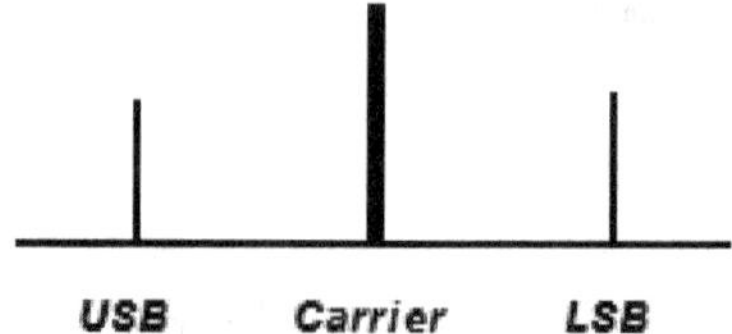

Figure 14.4 AM sidetones

<u>Problem</u>

Find the frequency of the upper and lower sidebands of an AM transmission on 570 kHz, if a 1kHz tone was being broadcast.

<u>Solution</u>

The upper sideband is at the carrier frequency + tone.

570 kHz + 1 kHz = 571 kHz

The lower sideband is at the carrier frequency – tone.

570 kHz – 1 kHz = 569 kHz

Typically, the sidebands will contain a range of frequencies when music or even voice is broadcast. The bandwidth is the entire area between the extremes of the sidebands. In our problem, the bandwidth of the transmission was 2 kHz. It may be necessary to limit the highest audio frequencies to the transmitter to stay within the FCC bandwidth allocations.

In an FM transmission, the wave's amplitude stays constant while the frequency changes according to the loudness. A louder signal equates to more of a *frequency deviation*. And, a higher frequency tone that is broadcast will cause the deviation swing to happen more rapidly. FM is a little harder to imagine than AM. It might help to picture a *Slinky* toy. A Slinky is a toy that is like a spring that varies in the compression along its length. A slinky in operation is similar to how a modulated FM wave will look. It is bunched up in some sections and elongated at other parts along its length. The frequency modulation process usually requires a great deal more bandwidth than AM. There is theoretically an infinite number of sidebands with FM. This result can be analyzed through a high level of mathematics called *Bessel functions*. In practice, it is possible to limit the amount of deviation to reduce bandwidth to acceptable limits. Many two-way radio systems use narrow-band FM, where the deviation is limited with the highest audio tones reduced so that the entire bandwidth of the signal is no larger than would be produced by an AM transmission. In narrow-band FM, the lack of high fidelity is not an issue because the voice range of frequencies does not go much higher than 3 kHz. The advantage of narrow-band FM is that it is not as susceptible to external noise as is AM.

The type of modulation that is used for digital modes utilizes a *phase shift* of the signal to convey information. A phase shift is a very slight change in frequency. A slight shift of frequency cannot contain much information, but this works fine for the binary system because each bit only contains only two possibilities. The *phase modulation* system, however, must be able to rapidly switch between the two phases to convey the intelligence information in packets of many bits.

Section 14.3. Resonance

When we discussed the inverting Op-amp, part of the "out of phase" output signal was fed back to the input. This signal is negative feedback and reduces the gain of the amplifier. As an example, there is both positive and negative feedback that you get as an employee. The *negative feedback* may be a reprimand for coming late to work. *Positive feedback,* such as complimenting an employee who does a good job, reinforces the behavior. The principle of resonance uses positive feedback as *constructive reinforcement* to continue an oscillation. As a continuous wave repeats the process every cycle, some energy from the input source is required to sustain the condition. If you were to tap a tuning fork, the sound would be noticeable for a while after the impact. Eventually, the sound would *dampen* out, and the fork would require another jolt to continue to make a tone. The tone that the fork makes is dependent on its physical structure. Generally, larger objects make lower tones than smaller objects, and this makes sense because we know that lower frequency waves have a longer (larger) wavelength. There is a case where a bridge in Washington State called the Tacoma Narrows Bridge was inadvertently first built with the dimensions that caused it to be resonant with the wind. During one storm, the wind energy caused constructive reinforcement to the point of vibrating the bridge apart.

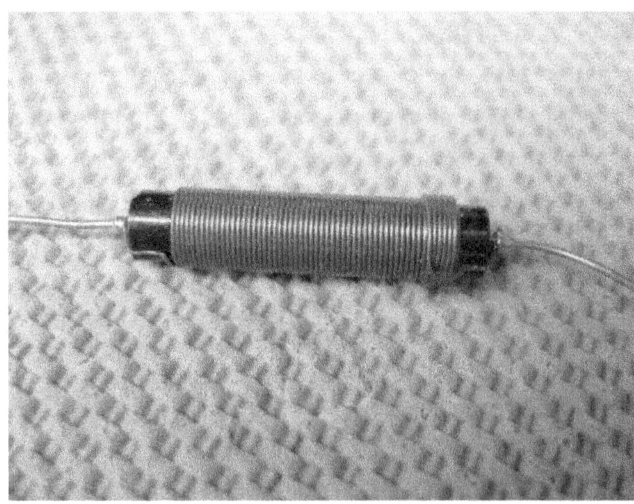

Figure 14.5. An Inductor, sometimes called a coil or choke

Resonance also takes place in electronic circuits, where the signal effectively vibrates back-and-forth between two components. The two electrical components exchange energy in almost the same way as mechanical energy makes a tuning fork vibrate. For an electrical circuit to vibrate as an electrical oscillation, you need to connect a capacitor to an inductor. As we discussed in an earlier chapter, a capacitor stores energy in the form of an electric field, and the inductor shown in Figure 14.5 is a coil of wire that stores energy in the form of a magnetic field.

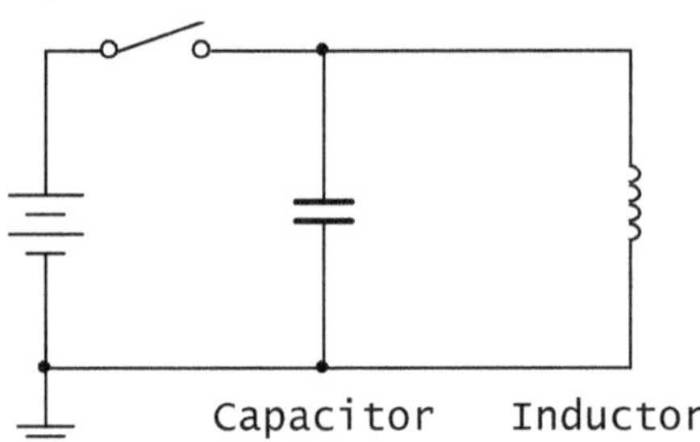

Figure 14.6 Parallel tank circuit

When the capacitor and inductor are parallel, as in Figure 14.6, it is called a tank circuit. When we close the switch contact, it is like one tap to a tuning fork. As pictured in Figure 14.7, the oscillation will dampen out as the circuit's internal resistance transforms the energy into heat. A small trickle of energy would be needed to sustain the oscillations.

Figure 14.7. Dampening wave

The physical structure of a tuning fork determines the frequency and wavelength of its tone. In a resonant electronic circuit, the frequency and wavelength are determined by the capacitance and inductance values. The capacitor is given the letter C, and the unit of capacitance is the *Farad*. The inductor is given the letter L, and the unit of inductance is called a *Henry*. (The letter I is already used to represent electric current.) A resonant condition exists when each component's resistance to AC called *reactance* is equal. The amount of reactance is expressed in Ohms. Capacitive reactance is given the term X_C , and X_L is used to represent inductive reactance. Reactance is an *imaginary number* because it is perpendicular to the real number line. Real resistance is on the real number line, while X_C is 90 degrees above it, and X_L is 90 degrees below it. The capacitive Ohms and the inductive Ohms are 180 degrees out of phase. No true energy is dissipated by reactance. To calculate the reactance in Ohms, the following formulas are used:

$$X_L = 2 \pi f L$$

$$\text{And } X_C = \frac{1}{2 \pi f C}$$

When $X_C = X_L$, the circuit is in resonance, and the reactive Ohms cancel out. All that is left are the resistive Ohms. We will derive the formula for resonance by putting the two reactance formulas equal to one another and solving for f:

$$2 \pi f L = \frac{1}{2 \pi f C}$$

$$f^2 = \frac{1}{4\pi^2 L C}$$

$$f = \frac{1}{2\pi \sqrt{L C}}$$

<u>Problem</u>

Find the frequency of resonance for a tank circuit, if a 10 pF capacitor is used with a 5 mH inductor.

<u>Solution</u>

p is the engineering prefix for pico, which is 1 x 10^{-12}, and m is the engineering prefix for milli, which is 1 x 10^{-3}.

$$f = \frac{1}{2\pi\sqrt{\left[\left(10\times10^{-12}\right)\left(5\times10^{-3}\right)\right]}}$$

$$f = \frac{1}{2\pi\sqrt{50\times10^{-15}}}$$

$$f = 712 \text{ kHz}$$

The tank circuit in the problem could be used as the frequency-determining circuit to generate oscillation in the AM broadcast frequency range.

Section 14.4. Transmitters

A CW transmitter is basically an oscillator with a power amplifier connected to an antenna. For AM, a chain of audio amplifiers mixes the audio with the RF carrier and then goes to an antenna. The mixing that occurs is also called *heterodyning* and must take place in a *non-linear device* like a tube or transistor. A resistor is a linear component because it can be shown to follow a relationship that forms a straight line on a graph, as in Figure 14.8.

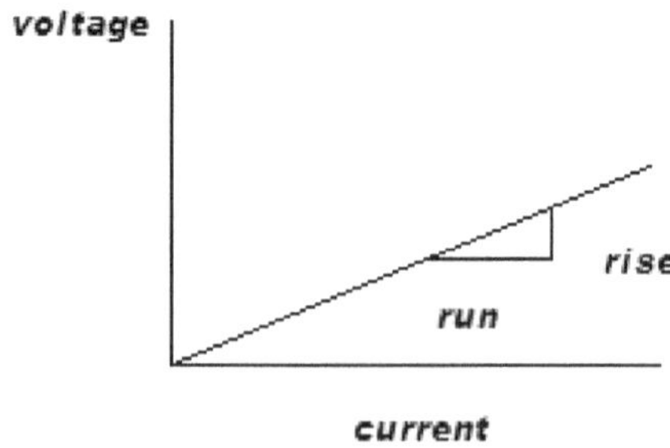

Figure 14.8.

156

When the voltage increases across a resistor, there is a corresponding increase in current through the resistor, as determined by the amount of resistance. The value of an unknown resistance could be found this way by using the slope of the line. Slope = rise/run. Mathematically, since the resistor follows Ohm's Law, and there are no exponential terms in the formula, it is a linear formula. Tubes and transistors do not follow a straight line. When the transistor's collector current is plotted against the base current, we find a curve, as shown in Figure 14.9. The transistor also reaches saturation, after which there is no additional increase in collector current with any increase in base current. There is a section that is the most linear, but for mixing purposes, we want to use a non-linear area.

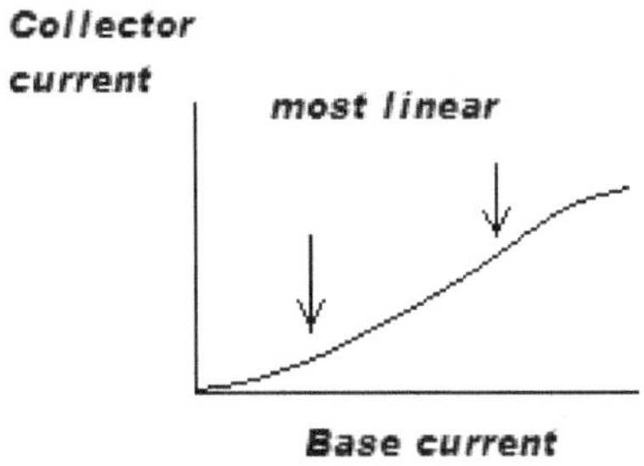

Figure 14.9 Transistor curve

Once the modulation process has taken place, any further amplification must be linear. The amplifying circuits would then need to be designed to operate in the characteristic curve's most linear area. Modulation that is completed before the final amplifier is considered to be low-level. If you modulate in the final amplifier, it is considered to be high-level modulation because the audio must be brought up to a high power level. High-level modulation offers the most efficiency since the high power amplifier can be operated on the low part of the characteristic curve and wouldn't be on all of the time. This type of operation would rely on the resonant tank circuit to provide oscillation through the exchange of energy between the capacitor and inductor. This is called the *flywheel effect*, and the amplifier's operation is considered class-C because it only needs to be on for a short amount of time for each wave.

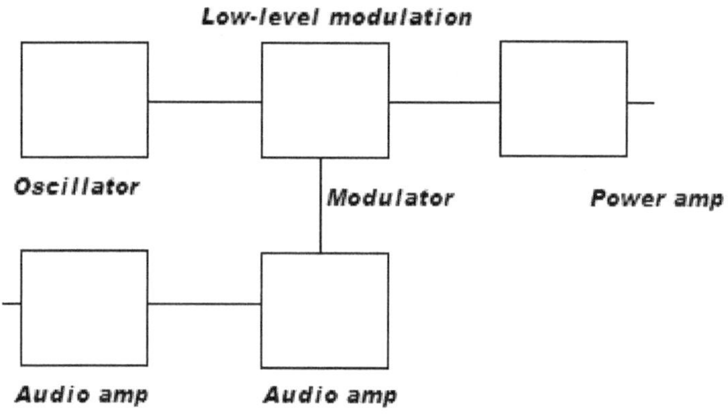

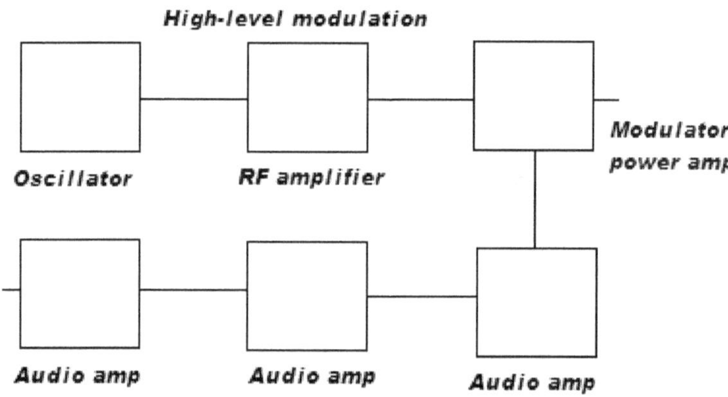

Figure 14.10 Block diagrams of low and high-level modulation

The high-level modulation scheme uses a linear power amplifier as the last stage in Figure 14.10. The low-level block diagrams represent Class C amplification with the power amplifier running in a nonlinear fashion to conserve electricity.

The newest AM transmitters use a Pulse Width Modulation (PWM) system, with pulses lengthened or shortened depending on the modulating signal's amplitude. The pulses are not broadcast directly but instead go through integrating circuitry to produce an AM envelope. PWM, is described in Figure 14.11; it is considered Class D and is very efficient.

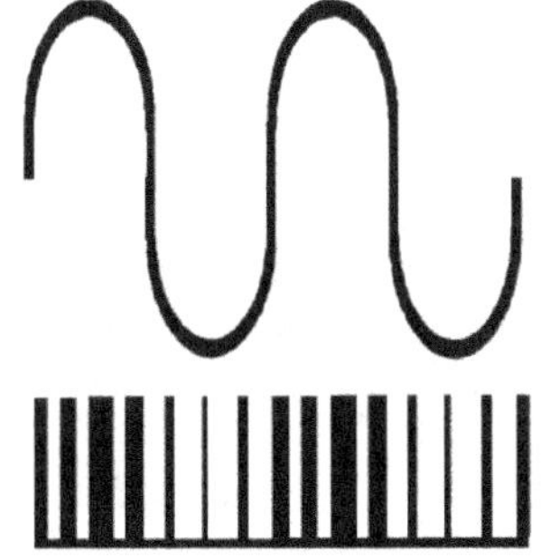

Figure 14.11. Audio to PWM

In FM transmitters, the audio is used to vary the carrier signal's frequency, and all power amplification must be linear. Some transmitters use a reactance modulator. There is a very slight amount of capacitance between the two sides of a diode's PN junction. This characteristic is utilized in a special diode called a *varactor* shown in Figure 14.12.

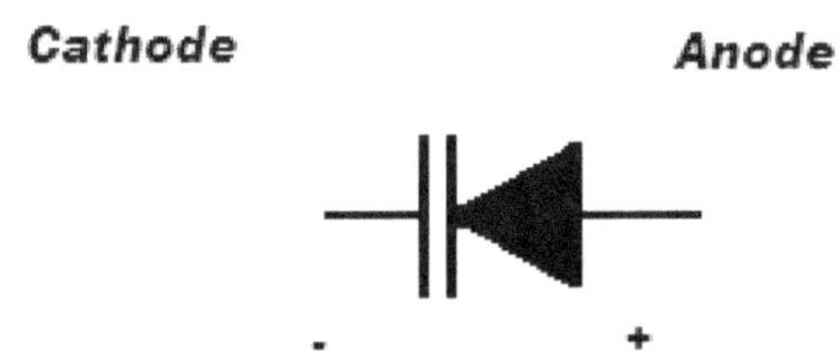

Figure 14.12. Varactor Diode. (The vertical lines signify capacitance.)

The varactor is placed across a tank circuit, and a reverse polarity DC voltage controls the distance between the two sides of the junction. With more reverse bias, the sides move farther apart, which lowers the capacitance. With less reverse bias, the side move closer, and the capacitance is increased. The varactor is varied with changes in audio so that the resonant frequency of the tank circuit - frequency modulates the carrier.

Section 14.5. Receivers

There are project kits called *crystal radios*, which can be constructed to tune in AM radio broadcasts. They are easy to build, and children have no problem constructing them. The crystal working in conjunction with a thin piece of metal called a *cat whisker* work as a rectifier, and we earlier learned that rectifiers turn AC into DC. Radio waves are a high-frequency AC signal. With AM broadcasts, the audio seems to ride the carrier wave. The rectifier removes the bottom part of the envelope waveform. A capacitor placed in parallel eliminates the remaining high-frequency carrier while leaving the lower frequency audio wave on the top half of the envelope. We will demonstrate the actions of receiving an AM signal and detecting the audio in the next few pictures. The AM modulated transmitted envelope signal is shown on an oscilloscope in the Figure:

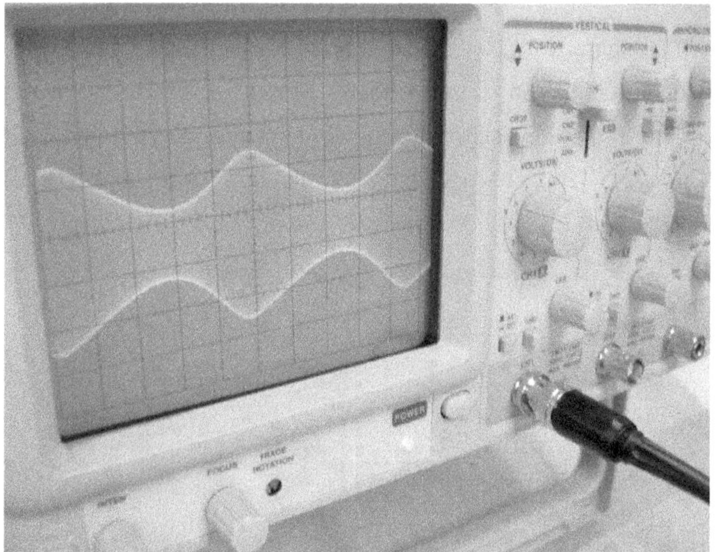

Figure 14.13. AM envelope

The cat whisker is replaced with a diode shown as part of the circuit of Figure 13.14. Usually, a diode made of germanium is used because it has a lower voltage drop.

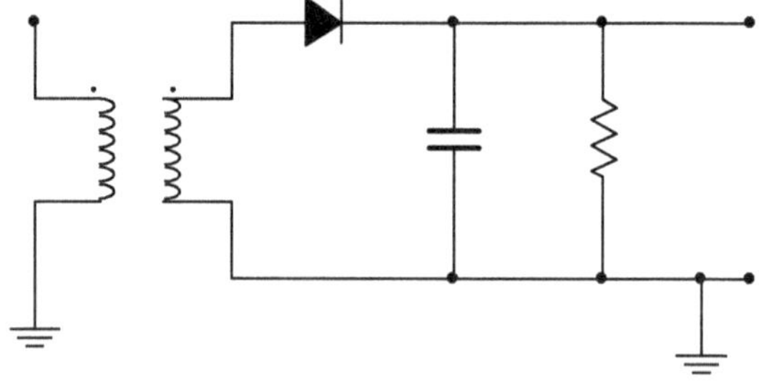

Figure 14.14 Demodulating circuit

160

After rectification takes place, the capacitor shunts the high RF frequency carrier to ground. All that is left is the audio wave developed across the resistor as shown on the oscilloscope in Figure 14.15. The final output is the sound broadcast from the radio station. We are using a 1 kHz tone to modulate the carrier in this example.

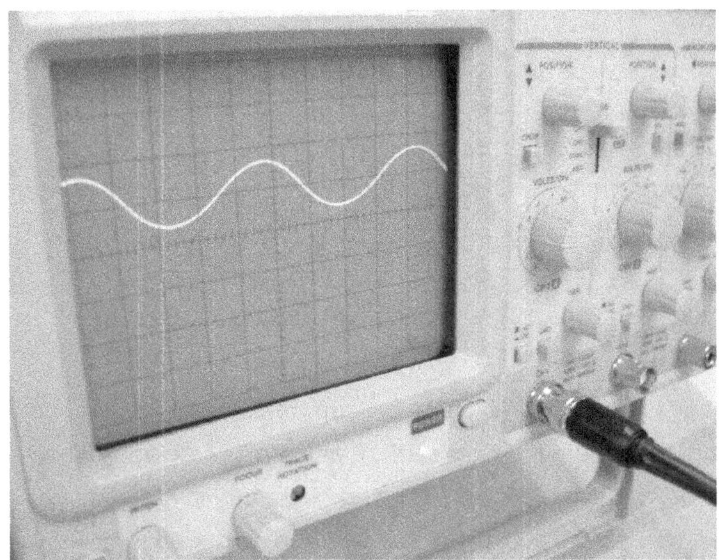

Figure 14.15. Recovered audio

In a crystal radio kit, there is usually no amplifier, and headphones must be used to hear the sound. If an audio amplifier were connected to this circuit, it would be similar to a normal AM radio receiver. The section that we just discussed is called a *detector* because it detects the audio. It is sometimes also called a demodulator. A transmitting radio station's frequency must be selected in a receiver. One method of accomplishing this is to feed the antenna signal across a tunable tank circuit. The listener can choose the frequency of the desired radio station by adjusting the resonance of the tank circuit. It is possible to use either a variable capacitor or inductor as the tuning component. This type of radio is fundamental and is called a Tuned Radio Frequency (TRF) device.

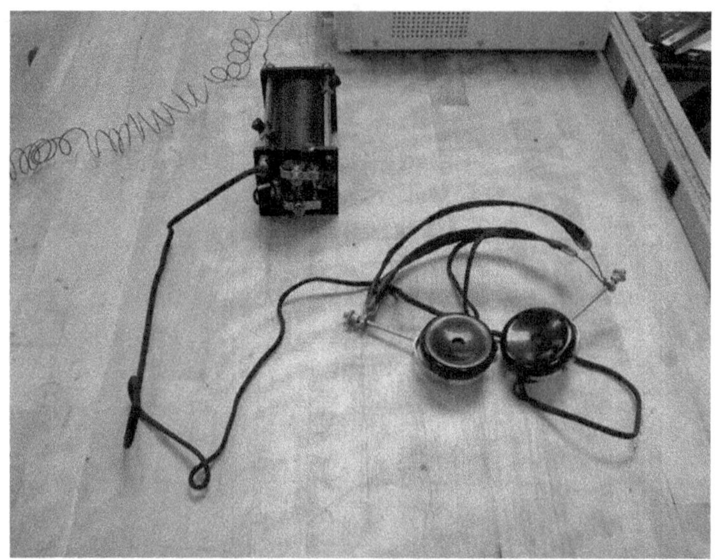

Figure 14.16. Actual 1930 era crystal radio

Selectivity and the ability to pick up weak signals have improved from our crystal radio set pictured in Figure 14.16 and the TRF design that we just examined with a *superheterodyne* receiver. The superheterodyne radio is what is commonly used today. Just as the audio is mixed with the carrier on a modulated transmission, the superheterodyne receiver mixes an internally generated tuning signal to develop a fixed intermediate frequency (IF). In standard AM broadcast radios, the IF is 455 kHz, and 10.7 mHz is used for FM radios. When mixing occurs, a sum and a difference frequency is generated. Those signals, along with the two original frequencies, are sent to a tank circuit that is resonant to the desired IF frequency. Usually, the difference between the radio station frequency and the tuning frequency - developed in the *local oscillator* (LO), is used as the IF. The IF stage has a number of high gain amplifier circuits to boost the signal before it is sent to a detector for *demodulation* and then sent to an audio amplifier. Figure 14.17 is the block diagram of a superhet receiver.

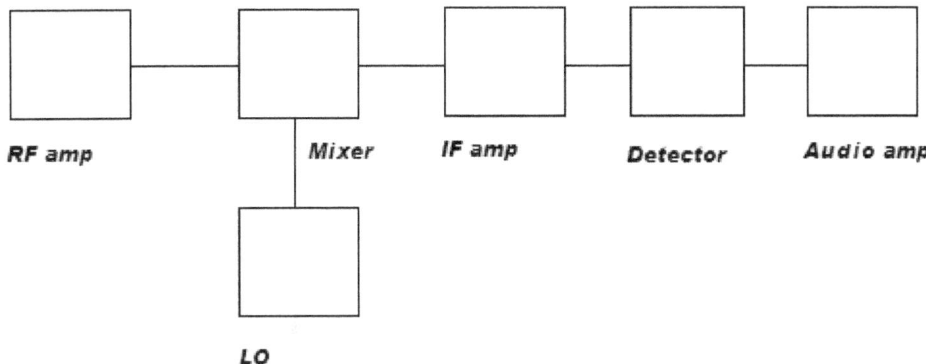

Figure 14.17. Super heterodyne receiver block diagram

In an FM receiver, the main difference is in the detection stage, where the changing frequency is interpreted as audio. One of the easiest ways to do this is with a *Phase Lock Loop* (PLL). A PLL is a versatile circuit that can be used to keep a transmitter from drifting off frequency by generating an error voltage used to correct the drift. In a receiver, the error voltage follows the shifting modulated carrier of an FM transmitter. The error voltage resembles the audio wave. The circuit is called a discriminator rather than a detector. The amount of frequency deviation from the carrier center frequency is related to the loudness of the audio. The rate of change of the deviation is related to the audio frequency. Analog broadcast FM is capable of reproducing audio up to about 15 kHz. Digital broadcasts can produce CD-quality sound up to 20 kHz. And, there may be more than one program on a digital broadcast, along with digital information such as the artist's name and song title. Additional decoding circuitry is incorporated into a digital radio receiver.

Chapter Fourteen summary

Radio frequencies are a part of the electromagnetic spectrum and are limited in availability. Governments grant broadcast licenses to utilize sections of bandwidth in different ranges. AM broadcasters use medium-wave channels while FM broadcasters use VHF channels of wider bandwidth. Cellphone services are in the upper ranges of MegaHertz up to the low GigaHertz range. 5G is at the highest edge of phone frequencies and offers the widest bandwidths. Hartley's Law states that information is proportional to bandwidth multiplied by time. FM radio broadcasters use much more bandwidth than AM broadcasters to produce high fidelity sound. There are specialized communications services that use a limited bandwidth called narrow-band FM. The radio frequency section of the EM spectrum contains many different radio services. To convey information with RF, the first and simplest method was to pulse the transmitter's carrier and send Morse Code. Many modulation techniques exist where the information is mixed with the carrier and then demodulated at the receiver. Radio frequencies use a process called resonance to build oscillations that propagate through space. The carrier usually does not contain information but is a transmit vehicle and mixed with the intelligence signal to be transported. There are many different types of modulation methods, but all must impress information on an analog carrier. Even digital transmissions must vary an analog component in some way since radio waves are analog.

Chapter 15

Computer networks

Section 15.1. Networks are for more than just television

The popularity of special media transmission services is cutting into the profits of the over-the-air broadcast networks. There are selections of just about anything you might want on cable, satellite, and streaming services. Some of the services have a worldwide reach and are viewed by hundreds of millions of people. In a broad sense, a network is a connection. Broadcasting is a one-way dissemination of information, but other types of networks allow for a two-way exchange. In the business world, people network by joining groups like the Chamber of Commerce or the Rotary. The members network with like-minded individuals and build relationships that may help their careers and businesses. Most computers are in some way networked with other equipment to exchange information. Thanks to the Internet, people can quickly access and exchange information worldwide. The Internet is a network of networks.

The very early computers were *mainframes*. There was a central computer connected to every user's keyboard and monitor. The central computer polled each *workstation* for a very brief period of time. The session was repeated after all other users were polled, so that each user felt like they were constantly connected. There was a minimal amount of workstation memory that acted as a buffer between connection periods. This process of sharing time is how some satellite TV and radio broadcasters maximize efficiency. The concept is called *Time Domain Multiple Access* (TDMA). When the PC came into use, each workstation had enough computing power to operate on its own, without a mainframe connection. The PC hardware and software on individual machines were able to accomplish many tasks without the need for a link to an external computer. When a file needed to be sent to another workstation, it was transferred by way of a floppy disk. However, in large offices, it became more efficient to begin connecting the machines, and networks were formed.

Two computers can be directly connected and can exchange baseband signals. One way this was done on old computers was by using the *serial port* as a *null modem*. The word *modem* is a conjunction of the words *modulate* and *demodulate*. Most modems convert between digital data and analog signals used for transmission. The word *null* means not, so two computers do not need a modem to communicate, but serial ports are no longer used and are a legacy connector. They looked very similar to a VGA monitor connector. The monitor connector has three rows of pins, but a serial port only had 9 pins in two rows. It is sometimes called a *DB9*. There are many different kinds of ports. On the ocean, ports are where ships dock to exchange people or cargo. In a computer, the ports exchange data, and there are both hardware and software ports.

A networked computer must have an actual connection. It could be a hardwired or wireless connection. The design of the network connections is referred to as the *topology*. Also, for the network to function, it must have *protocols*. A protocol is an agreement between parties. If nations worldwide agree to behave in a certain way, they will sign off on a protocol. On a computer network, protocols dictate how computers will behave when communicating. The most popular style of hardwired baseband network connection today uses a bus or star topology called *Ethernet*. With Ethernet, the connectors look similar to telephone jacks and are known as RJ-45 connectors. They are a little larger than a standard telephone jack and connect twisted pairs of wire called Category-5 (Cat-5) or Cat-6. Some computers may have a *Network Interface Card* (NIC), which is a circuit card located perpendicular to the motherboard. Most motherboards now have the NIC built-in. The network interface has a unique address known as a *MAC*. It is a 48-bit code number assigned during manufacture that specifically identifies the computer for network purposes. The computer listens for its MAC address when connected to a network and accepts network traffic sent following the transmission of its address. Otherwise, it ignores data on the network. When a computer on Ethernet wants to send data, it just burps it out on the network. If another computer happens to transmit at the same time, it is called a collision. Both computers will detect a collision of data and will each wait for a random period before transmitting the data again. There can only be one signal on a baseband bus at any given time. The Ethernet system may seem a little haphazard, but it is extremely fast, even with the collisions of data.

Section 15.2. Intranets and the Internet

Computers, networks, and the Internet intimidate many people. It is easy to be intimidated because of the powerful nature of the vast information exchange provided through computers on the Internet. Modern computer hardware and software operations can be so amazing that it may be a bit overwhelming. Many so-called computer experts rattle off acronyms that they probably don't really understand. Do not be intimidated! There are tremendous amounts of fabricated terms in computer jargon, but they make sense if you take a few minutes to think about it. The underlying principles are the things to really try to comprehend.

When computers are networked together in an office or school, they are on a *Local Area Network* (LAN). A LAN is considered an *Intranet*. The Intranet is an in-house version of the *Internet*. (Notice the slight difference in the spelling.) You may be able to access the *Internet* from an *Intranet*. Some Intranets connect far-flung locations of a business. This type of structure can be called a *Wide Area Network* (WAN). Large corporations sometimes lease data lines from the phone company to securely carry the network traffic, but a more cost-effective method that is gaining in popularity is a *Virtual Private Network* (VPN). A VPN tunnels through the Internet to carry secure data. Security is a major issue when developing a professional network. Care must be taken not to allow unauthorized persons to access sensitive information. Wireless networks are especially vulnerable to attack. A wireless network uses RF signals, and the data can be picked up by anyone with a receiver. *Wi-Fi* uses short-range transceivers, but someone could eavesdrop from a greater distance if they use a sensitive receiver and high gain antenna. There is an encryption security feature on Wi-Fi units called *WEP* that only permits authorized computers to connect to the wireless network. In professional applications, some systems use more secure encryption than WEP. The WEP system uses a code that doesn't change, whereas more secure systems keep changing the encryption code. Firewalls are sometimes used to monitor incoming and outgoing traffic and block unauthorized access. *Firewalls* can be both hardware and software-based. Network communication over fiber optic cable is not only very fast, but it is almost impossible for any eavesdropping to take place without splicing into the cable. All of the signals remain inside the cable and only transfer from end-to-end. The process is called *Total Internal Reflection,* and no signals leak from the sides of the cable. Fiber bundles, shown in Figure 15.1, contain glass cables smaller than the size of human hair and are very difficult to splice together without specialized equipment. The most secure computer is not connected to a network, but that is also the least functional.

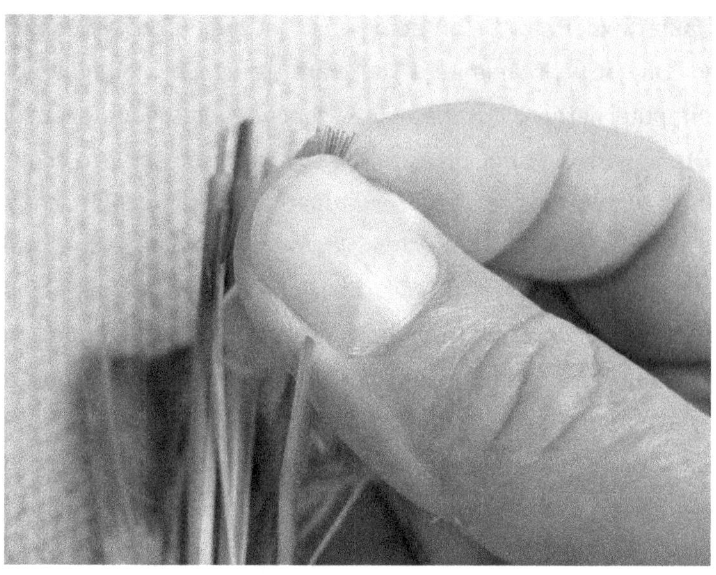

Figure 15.1. Twelve fiber optic cables in one bundle

A Computer can connect to another computer on the network through a hub or switch, but generally, a workstation will connect to a server. A server is a computer that is dedicated to providing services such as transmitting Web pages, handling emails, or storing data. A server has high-end hardware and a *Network Operating System* (NOS). Good servers also have redundant hardware and a back-up source of power for reliability. Most Intranets connect workstations to servers through *switches* that make as many of the short-range local connections as possible, to minimize data flow on the *backbone*. A backbone is the main link that may run throughout an entire building. The Internet connects local networks together through a system of routers. *Routers* are devices similar to switches but connect computers on different networks. That is why we say that the Internet is a network of networks.

The high level of technology that we have today all began with the human race's desire to communicate over great distances. Before electric power existed and electronic devices were developed, civilizations used novel means to communicate. They used smoke signals and noises like the sounds from drums. In ancient times, people ran between two points to deliver messages or traveled on horseback. The first technological way to communicate over great distances was developed in France during the time of Napoleon. They developed a mechanical way of passing information between towers separated by great distances. Observing visual indications from one tower to the next enabled the operators to pass the messages along. Depending on weather conditions, the information could take days to go across the country. Then in America, we had the invention of the telegraph and the telephone. As we know it, the Internet came about from the idea of using a *packet switching* technique. Up until that time, connections were continuous for the entire length of the communication. With packet switching, the information is broken up into small packages and delivered by the fastest means possible. Packet switching is how the Internet functions today. Small packets take the best route from point to point. The military designed the Internet to be redundant. So, if one path became unusable, an alternate path could be taken for the communication to get through. The data is also checked, so that missing packets may be resent. We find that many times a system initially designed for military use winds up becoming an invaluable asset for peaceful purposes. Along with the Internet, the Global Positioning System (GPS) is a great example of a concept that had military beginnings but now is utilized mainly for peaceful civilian purposes.

Every device connected to the Internet has a unique 32-bit number assigned for routing purposes. The largest number in decimal would be 255.255.255.255. The leftmost side of the number identifies the *Internet Service Provider* (ISP), and the remaining numbers identify the customer of the ISP. They assign numbers in two ways, a static number never changes, and a dynamic number changes every time a customer logs in to the ISP. The ISP keeps records of the customer/number association at a specific time of day, so it is tough to be truly anonymous while online. The numbers are the address for your computer connection. When you want to look at a Web page and access a server, it needs to know where to send the data. Everyone is a number on the Internet, even servers. If you want to see the latest model Ford trucks and type "www.ford.com," specialized computers on the Web called Domain Name Servers (DNS) will convert the letters to numbers in the 255.255.255.255 scheme. The numbing that we have just discussed is *Internet Protocol* (IP), and the packet transfer system that we discussed earlier in this section is called *Transfer Control Protocol* (TCP). Together they are TCP/IP, and that is the way that the Internet is structured. The current 32-bit IP addressing scheme can generate approximately 4 billion numbers. Plans are underway to add additional bits to accommodate more Internet users and IoT devices. IPv6 uses 128 address bits and is in limited use; however, it is not backward compatible with 32-bit addressing.

Chapter Fifteen summary

A stand-alone computer is the safest and stable but also not as functional as having the ability to draw on outside resources. Computers can be connected in a peer-to-peer fashion using wireless or wireline connections. Many of the wireless considerations were covered in Chapter Fourteen. Fiber optic and Ethernet are the most common methods for hardwired connections. Ethernet uses baseband signals transmitted on Cat-5 or Cat-6 cables using RJ-45 connectors. The topology may be bus if computers are connected to a hub. Bus topology has packet collisions as a normal part of the exchange of data. A star topology can be implemented by using switches, where each machine is connected to a port on the switch, and intelligent switches can then route data exchanges between specific computers. A MAC address is a physical hardware descriptor, and IP addresses are logical designations. Hubs and switches are useful in keeping local traffic off of the main backbone. A local Intranet network is called a Local Area Network LAN, and if it incorporates a wider geographical area, it is called a Wide Area Network WAN. Servers may be a part of an Intranet. Routers can connect to the Internet, where TCP/IP protocol is used. Domain Name Servicers DNS translate Web addresses to IP addresses and forward traffic. Currently, 32-bit addresses are being used, but with the proliferation of computers, phones, and IoT devices requiring Internet connectivity, a new 128-bit addressing scheme called IPv6 has been introduced.

Chapter 16

Green technology and innovation

Section 16.1. Green technology

It makes sense to design products that do not harm society, but there are always tradeoffs as we progress technologically. As industry developed in the mid-twentieth century, it became apparent that with significant developments in technology, many unwanted consequences came. Cancer-causing agents like PCB's in electronic components, mercury, and other heavy metals were routinely used in industry and discharged into wastewaters, and soot from smokestacks fouled the air. Government regulations balanced the need to protect people's health and the environment with the need for a vibrant economy. Improvements in education and social awareness of people becoming interconnected helped our nation form an environmental consciousness. Many developing countries worldwide are still grappling with health, safety, and pollution and balancing them against economic growth.

Creating products and using technology that is friendly to the environment began to take on added importance, starting in the 1960s. Earth day began in 1970 as a way to promote the subject of *sustainability* and environmental preservation. The European Union has led the way with the adoption of sustainable products and strict regulations. They were a driving force to remove lead from electronic solder, and in some cases, even require that manufacturers have a recycling process at the end of a product's life cycle. If done correctly, recycling is good for the environment as well as for the bottom line. Many times, precious metals like gold and silver can be recovered from recycled products, and other more common elements can be recovered as well. Nucor Steel Company is an excellent example of profitable recycling. They use an electric heating process to melt scrap metal to produce new steel products. Great strides in green technology have been made in the last few decades, but there is much more work to be done. The human race coexists with nature, and it is important always to try to minimize any adverse impacts of technology.

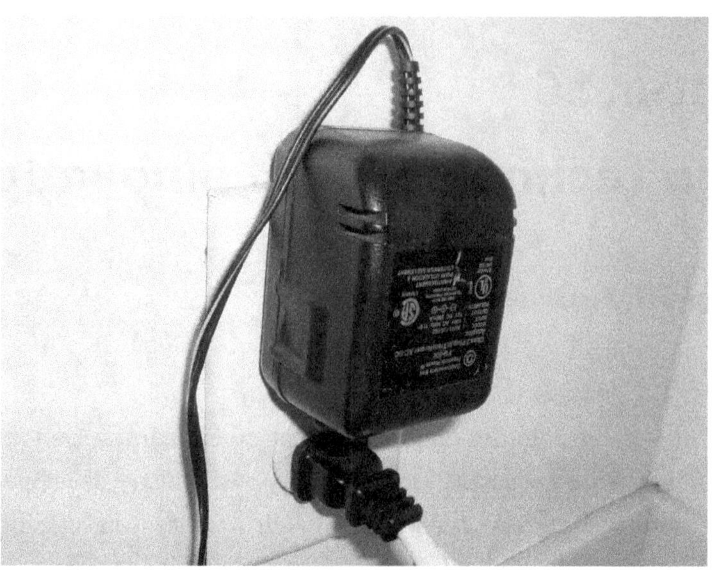

Figure 16.1 Portable power supply

It makes sense that we can lessen any detrimental effects that technology might have on the environment by minimizing waste. Transformers inside of power supplies, like that of Figure 16.1, are used to operate and charge small electronic devices but waste energy. These power supplies are sometimes called *wall warts.* The transformers produce heat even when the device is off and not charging. The heat is a result of a nonproductive current flow that occurs inside the core of the transformer called an *eddy current.* It is like an eddy current that occurs in a river, where the water continuously circles and doesn't flow downstream. The newer power supplies, like those used in computers, have a better method of voltage conversion and are called *switching* power supplies. They convert the 60 Hertz AC to a higher frequency and process the energy conversion more efficiently in transformer cores that are not affected by large eddy currents. You can save energy by unplugging your warts when they are not in use, and whenever possible, use a switching supply. It's good for the environment, and it will help to lower your electric bill.

The need for alternate energy has taken on added importance as climate change has become increasingly evident. We have been getting energy from stainable sources like water for a long time. In a *hydroelectric* plant, the moving water spins a turbine that produces electricity. Ingenious concepts are also being examined, such as using the ocean's water currents or tides to produce energy. The wind is used to produce electric power both on and off-shore. The photovoltaic process captures energy from the sun, and the efficiency of solar cells, which averages only about 20% now, is expected to be over 50% within a few years.

Eliminating fossil fuel-powered vehicles as electric power gains traction from innovations in battery technology is becoming more widely accepted. Green technology is providing employment opportunities and the promise of a healthy future for our world.

Section 16.2. New product development

Humans are a progressive species. We develop very powerful tools and shape the world around us. The fact that we innovate might be programmed into our DNA. It is as if a physiological trait has us evolve the world to fit our growth, rather than us changing to fit the world. When our ancestors were in caves, they painted and drew on the walls. Above ground, they formed towns and nations. In the agrarian age, people built dams and moved water for irrigation, and enriched the soil with nutrients. Plants were modified through selective breeding to produce optimal crops. Pests were either eradicated or discouraged from causing damage to planted areas. As the human race moved into the industrial age, changes to the world became even more pronounced and far-reaching. Some believe that we have gone too far in changing the world and may have damaged it beyond repair with the looming specter of global warming. The doomsday threat of the cold war era has returned to haunt us, and we could destroy the entire world through nuclear war. Comprehensive knowledge is increasing at a phenomenal speed, and we can only hope that our wisdom is not too far behind. As we moved from the industrial age into the information age, we have products and services that were only things dreamt about a few years ago.

There are two types of product development. The first type of development occurs quite regularly and is called *incremental* development. When a product is updated and slightly changed, it is said to be incrementally improved. In some products, it is expected that models be updated frequently. The automobile companies spend a vast amount of money making only minor changes to body styles each year. The newer model car may only be slightly improved, but if the manufacturers failed to make any changes, their sales would suffer because people expect progress, whether real or perceived. The other type of development sounds bad, but it is not. A *disruptive* product development may completely change the way things are done. The initial development of the automobile was a disruptive development. It drastically changed society and put many people in the horse and buggy industry out of business. Disruptive developments don't usually happen too often. A more recent example is the development of home personal computers. Apple Computer co-founder and visionary Steven Wozniak was an electronic and computer hobbyist who was instrumental in designing and building the first viable home computer. He wanted the home computer to help people do things better. It revolutionized the way the world works by increasing productivity. For a long time, America had a competitive advantage in business over other nations of the world. We had access to high-tech manufacturing techniques that companies in other countries weren't able to afford readily. We could produce high value-added products and services and sell them at a premium, but with mass production and lower costs for hardware and software, the playing field has been leveled. The traditional computer has been incrementally developed into a phone. Many products that began as a disruptive development soon became commonplace. Some might consider virtual reality (VR) to be a disruptive innovation since it incorporates human movement into a game environment and changed the way video games are played, and can also be used for much more than just entertainment. Electric vehicles and alternate energy may also prove to be a disruptive innovation. Work is being done on devices that may revolutionize mechanical products. Micro Electro Mechanical Systems (MEMS) uses IC fabrication technology to construct miniature robotic machines that are as small as 1 millimeter, 39 thousandths of an inch. These devices are considered to be a system because they contain a CPU, sensors, and actuators. These devices may even be programmed to construct other miniature devices.

It once was the case that a single person working alone in a garage or basement could come up with the next great invention. While that still may be true, it is far more likely that a team of people will be needed to develop a high technology product. Because of the increased complexity of technology, many individuals may need to work together to bring an idea to reality. When Microsoft began writing their first programming code called *Basic* in the 1970s, they only had a few people working on the project. Their latest operating system contains over 50 million lines of code. It takes thousands of programmers working in development teams to bring that kind of project to fruition.

Today in product development, interdisciplinary teams work simultaneously on most large projects. Experience has shown that cross-functional teams perform far better than organizations that keep their workers separated. The most successful projects have seemingly unrelated functions such as manufacturing, design engineering, accounting, marketing, finance, and others all working together from the inception. This system provides efficiency and helps to identify any roadblocks that might materialize along the way. Designers may come up with ideas, but it is essential that they work together with a team of people who have different skill sets. As the saying goes, *the whole is greater than the sum of its parts.*

Section 16.3. Project management

If you live where it snows in the winter, you can appreciate the importance of a good snowplow. A project manager should act like a snowplow as a project moves ahead. They always need to look ahead and anticipate potential problems that may arise and push things along.

Figure 16.2 Building construction

We tend to only associate project management with civil engineering projects like the construction of roads and buildings like in Figure 16.2, but project management is an essential part of new product development. The project manager is not a foreman but acts almost like a construction contractor. They must arrange for as many different functions to occur simultaneously as possible to maximize efficiency and ensure that the target date for completion is kept intact. There is a good visual tool to keep track of many different functions that are scheduled. It is a timeline called a *Gantt chart*. The chart lists operations horizontally on a day of the week chart. The chart is a visual way to determine the critical path for the progression of a project by showing simultaneous working activities and identifying those activities that have predecessor activities that must be accomplished before the new task can begin. Microsoft offers a program called *Project* that can generate Gantt charts and assist in other project management tasks. The manager will also need to identify *slack* time when work can not be done simultaneously, and will then redistribute resources during those periods to ensure efficiency.

A project manager is also a motivator and communicator and needs to have good people skills. They need to interact with all levels of an organization, as well as with contract workers. Continual monitoring of the progression of the project is performed. As goals are met, it can be thought of as the project passing through a series of gates. This type of benchmarking identifies any problems needing to be resolved, or that the project's progress is running on track.

Project management should be involved at the inception of a project. The manager can help to establish a realistic timeframe and cost projection. Once a project starts, *scope creep* must be minimized. Scope creep is when additional features are added to a product and new objectives are added to a project while it is in progress. This can prove to be detrimental to both the timeline and the budget. Extra time spent in preliminary planning and having a firm concept in place before beginning a project will drastically increase the chance of success and minimize stress.

Many new products can be conceptualized through brainstorming. The process could entail listing a number of ideas without making any practical judgments. After the list is established, the field can be narrowed to the most likely product idea. This is when a cross-functional team can really help. Ideas for new products are important, but it is the innovation that changes the world. Many times the words, *creativity* and *innovation* are used interchangeably. They are not the same. The difference is that creativity is *dreaming* and innovation is *doing*. To move forward, we need both. Dream and come up with novel ideas, and then make the dreams come true with hard work.

Chapter Sixteen summary

Sustainability is recognized as essential for both the health and viability of society. There is always a trade-off between environmental protection and economic development. The line of separation has become increasingly blurred since environmentally sound practices and procedures, can generate jobs and increase financial well-being. New products require an inspirational concept, but it is innovation that brings an idea to fruition. Project management is best accomplished by employing cross-functional teams working concurrently toward the same goal.

Appendix
Review of mathematics

Algebra uses letters to generalize arithmetic operations. The same rules that apply to numbers also apply to the letters of the alphabet. We will review some of the most important properties and conventions.

Relationships and equations

A relationship shows how quantities relate to one and other and may not have an exact solution. An equation, however, is in balance. You can perform operations such as addition, subtraction, multiplication, and so on. To keep the equation in balance, whatever is done to one side must be done to the other side.

Conventions

The order of operation precedence is that multiplication and division must be done before addition and subtraction.

For a × b + c, the a and b must be multiplied, and the result is added to c.

When working with nested parentheses, always do the inside operations first. Handle the innermost parentheses first and work outward. In dealing with fractions, handle the numerator and denominator as though they were each inside of parentheses.

General Properties

Reflective a = a

Symmetric a = b then b = a

Substitution If a = b, then b can be substituted anywhere there is an "a" and vice versa.

Commutative a + b = b + a

Associative a + (b + c) = (a + b) + c

Distributive (a + b) c = a c + b c

Identity a (1) = a, and $\dfrac{a}{a} = 1$

<u>Properties of exponents</u>

$$x^0 = 1$$

$$x^1 = x$$

x^a is x multiplied by itself "a" number of times.

$$x^{-a} = \frac{1}{x^a}$$

$$x^a \, x^b = x^{a+b}$$

$$\frac{x^a}{x^b} = x^{a-b}$$